ANIMAL LIFE AT LOW TEMPERATURE

ANIMAL LIFE AT LOW TEMPERATURE

JOHN DAVENPORT

Professor of Marine Biology
University Marine Biological Station
Millport
Isle of Cumbrae
Scotland

CHAPMAN & HALL

London · New York · Tokyo · Melbourne · Madras

Published by Chapman & Hall, 2–6 Boundary Row, London SE1 8HN

Chapman & Hall, 2–6 Boundary Row, London SE1 8HN

Chapman & Hall, 29 West 35th Street, New York NY10001, USA

Chapman & Hall Japan, Thomson Publishing Japan, Hirakawacho Nemoto Building, 7F, 1-7-11 Hirakawa-cho, Chiyoda-ku, Tokyo 102, Japan

Chapman and Hall Australia, Thomas Nelson Australia, 102 Dodds Street, South Melbourne, Victoria 3205, Australia

Chapman and Hall India, R. Seshadri, 32 Second Main Road, CIT East, Madras 600 035, India

First edition 1992

© 1992 Chapman & Hall

Typeset in 10/12pt Plantin by Keyboard Services, Luton
Printed in Great Britain by T.J. Press Ltd, Padstow

ISBN 0 412 40350 1

A catalogue record for this book is available from the British Library

Library of Congress Cataloging-in-Publication data available

∞ Printed on permanent acid-free text paper, manufactured in accordance with proposed ANSI/NISO Z 39.48-199X and ANSI Z 39.48-1984

For Julia, Emma and Kate

Contents

Preface

To humans, cold has a distinctly positive quality. 'Frostbite', 'a nip in the air', 'biting cold', all express the concept of cold as an entity which attacks the body, numbing and damaging it in the process. Probably the richness of descriptive English in this area stems from the early experiences of a group of essentially tropical apes, making their living on a cold and windswept island group half-way between the Equator and the Arctic.

During a scientific education we soon learn that there is no such thing as cold, only an absence of heat. Cold does not invade us; heat simply deserts. Later still we come to appreciate that temperature is a reflection of kinetic energy, and that the quantity of kinetic energy in a system is determined by the speed of molecular movement. Despite this realization, it is difficult to abandon the sensible prejudices of palaeolithic *Homo sapiens* shivering in his huts and caves. For example; appreciating that a polar bear is probably as comfortable when swimming from ice floe to ice floe as we are when swimming in the summer Mediterranean is not easy; understanding the thermal sensations of a 'cold-blooded' earthworm virtually impossible. We must always be wary of an anthropocentric attitude when considering the effects of cold on other species.

Any book concerned with the effects of cold has to address the question 'what does cold mean?' On a temperature scale running from the near absolute zero of deep space to the million-degree incandescence of the surface of a sun, the difference between Antarctic winter and Saudi summer is quite insignificant, but to us it matters considerably. In many ways the question is unanswerable. A cryobiologist will choose temperatures below 0°C, since water is prone to phase changes at temperatures below its melting point. A mammalian physiologist will talk of critical temperatures below which a given mammal species has to increase its metabolic rate to maintain core temperature. A specialist in invertebrate biology may think in terms of temperatures which are low enough to prevent complete metabolic acclimation. For closely related species, the definition of cold may be quite different. Cold is a low temperature stress, a set of circumstances which demand adaptations at the species level, and responses at the individual level.

This book is concerned with the ways in which animals are adapted to deal with cold and is written from the standpoint of an ecophysiologist with strong interests in adaptive behaviour. Temperature effects are crucial to so many aspects of biology that the quantity of scientific literature devoted to cold is so great that exhaustive review is no longer feasible in a reasonable sized volume. The intention here is to provide a framework and basis for the study of the effects of cold by undergraduate and postgraduate biologists.

Chapter one defines a number of basic terms and concepts; particular emphasis is given to acclimation/adaptation, and to the biochemistry of heat production. Chapter two describes the cold environment, both as it is now, and as we judge it was in the past from current palaeoclimatological evidence. Chapter three is devoted to the behavioural mechanisms by which animals cope with cold, while Chapter four describes the anatomical and physiological adaptations which allow endothermic animals to maintain high core temperatures in a hostile environment. Chapter five considers sleep, torpor and hibernation in both ectotherms and endotherms. Chapters six and seven are both concerned with the biophysics and biochemistry of survival at subzero temperatures. Chapter eight describes the relationship between Man and cold environments, together with an outline of the technology he has developed first to combat cold, and latterly to use it to his benefit. Chapter nine attempts to evaluate the importance of cold to evolutionary processes; not in terms of its function as an agent of selection (which function is implicit in the evolution of the behavioural, physiological and biochemical adaptations described in the rest of the book), but in terms of the direct effects of cold on mutation and recombination rates, and in isolating populations so that allopatric speciation can proceed.

John Davenport
Isle of Cumbrae

PART ONE

Introductory Material

1 Basic concepts

1.1 TEMPERATURE

The temperature of a system is determined by the quantity of kinetic energy within it. In turn, the quantity of kinetic energy is a measure of molecular motion. The SI unit of temperature or temperature change is the Kelvin (K), now universally used in physics and chemistry. However, in the biological literature, where a relatively narrow temperature is important, there has been little movement towards use of the Kelvin (except for cryobiologists who eschew negative numerals), probably because it results in rather cumbersome numbers.

Throughout this book temperatures will be given in degrees Celsius, but in two forms. Actual temperatures will be indicated as '°C'. For example: 'the mean daytime rockpool temperature in June was 19.6°C'. Temperature differences will be given as a deg. C. For example: 'the air temperature fluctuated by 8 deg. C'.

1.2 WATER AND LOW TEMPERATURE

Water makes up most of the volume of animal tissues (usually between 60 and 90%), so much of the biophysics of cold is concerned with the energy state of water. Water is an unusual liquid; it has a great heat capacity (4.2×10^3 Joules kg^{-1} deg. C^{-1}) which allows the storage of vast quantities of heat in lakes and oceans, thereby buffering the sharp temperature fluctuations of the atmosphere. Water rapidly absorbs the extreme ends of the spectrum of the sun's radiant energy. Infra-red wavelengths are totally absorbed within a few metres of the water surface, limiting the warming effects of sunny weather to surface layers. Water also has its maximum density at a temperature (4°C) well above its freezing point (0°C). This is an uncommon attribute of liquids, but is crucial to animal life in fresh water (which is usually fairly shallow). Because of the temperature difference between freezing point and maximum density, ice formation generally occurs at the surface of bodies of water, and the deepest water remains unfrozen at a temperature of 4°C. This situation allows liquid water to be covered by ice quite quickly (which reduces the thermal exchange between atmosphere and water); the ice melts rapidly too when atmospheric temperatures rise. If water, like most liquids, was densest at its freezing point, convective action would prevent freezing until the whole body of fluid was at the point of freezing, which would occur suddenly and totally from the top to the bottom of the liquid column! The situation of sea water is rather different; sea water melts at about −1.9°C and has its maximum density at this tempera-ture. Freezing of sea water is only accomplished in the presence of very low air temperatures at high latitudes (usually at night). It should be remembered that much of the bulk of ice seen at the poles is not sea ice, but icebergs calved from glaciers, themselves formed from compacted snow. Once sea ice has formed it has a density substantially lower than sea water so it floats.

Ice, the solid phase of water, can exist in nine stable crystalline forms (Hobbs, 1974; Hirsch and Holzaphel, 1981). However, for most biological purposes we need only to consider one polymorph, Ice Ih, the 'normal' hexagonal lattice structure produced by hydrogen bonding (Figure 1.1). In solid ice the lattice is regular, with O–O distances of 0.275 nm and tetrahedral H–O–H angles of 109°38′. X-ray diffraction and neutron-scattering studies have revealed that a distorted lattice structure of this type persists in liquid water (although the term structure in a liquid refers to the position of moving

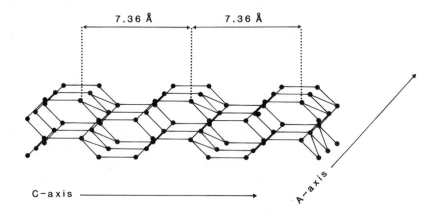

Figure 1.1 Diagram of structure of 'normal' ice (hexagonal ice, Ice Ih). Only carbon atoms are shown.

molecules, averaged over time — a probabilistic concept analogous to the Heisenberg uncertainty principle). When liquid water is cooled it reaches a phase equilibrium temperature at 0°C. This means that ice and water can coexist; it does not mean that ice will automatically form in liquid water at this temperature. In fact, water rarely freezes at 0°C, instead it becomes 'undercooled' at temperatures below 0°C. Crystallization (or condensation) of ice requires the presence of a group of molecules (known as a nucleus or, in older literature, a seed) onto which the molecules of water can condense. The ideal nucleus resembles the structure of the solid phase (i.e. a regular hexagonal lattice). Such nuclei arise spontaneously in pure water at all temperatures, but have a limited lifetime as molecular diffusion disturbs the configuration. However, as the temperature falls and molecular movement slows, such nuclei tend to become larger (thus having a greater surface area for condensation to take place), and longer lasting. The probability of crystallization/condensation therefore rises as the temperature falls. In pure water at normal atmospheric pressure the probability eventually reaches unity (i.e. freezing becomes certain) at a temperature of roughly -40°C, the temperature of homogenous nucleation (abbreviated at T_h), so this is normally the limit of undercooling.

Most water contains impurities ('motes'), particularly microcrystalline dust particles. These adsorb water molecules onto their surfaces, creating a large surface area for nucleation; they act as catalysts of freezing. The formation of such nuclei is known as heterogenous nucleation. The temperature at which nucleation (and hence freezing) occurs is abbreviated as T_{het}. $T_{het} > T_h$, and T_{het} is variable — it depends on the type and size of mote (see Franks, 1985 for a fuller discussion).

Ice crystals assume a variety of forms (Hobbs, 1974). Small ice crystals are formed if water is allowed to undercool to a considerable extent; large ice crystals are formed if liquid water is seeded or nucleated (e.g. by adding ice crystals) at relatively high temperatures. If small-crystal ice is warmed, the small crystals are gradually replaced with large ones (a process known as ripening). The most common form of ice crystal has a tree-like dendritic structure. Dendrites form because growing crystals lose their latent heat of crystallization to the liquid phase. Heat transfer rate (and hence growth rate) depends on the area of crystal in contact with the liquid. This area is greatest in areas of the crystal structure which are away from neighbouring crystalline areas. In consequence, crystals tend to grow in branching needles, away from one another, though usually there is a tendency for dendrites to grow parallel to temperature gradients.

The delivery of latent heat of crystallization into the liquid phase explains why undercooled water soon warms to the melting point (Figure 1.2) once freezing starts. Dendrites continue to grow until they touch, but nucleation is inhibited as the temperature rises to the melting point of water. At this stage

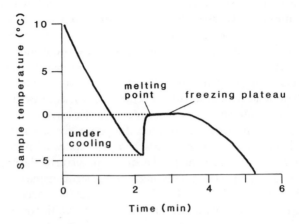

Figure 1.2 Profile of temperature change of a cooled sample of distilled water.

only about 6% of the water is frozen. Further loss of latent heat (and further formation of ice) is achieved by conduction of heat through the ice crystals themselves towards the nearest heat sink (e.g. cold air). Larger cellular dendrites are then produced. The slower cooling rate of water after freezing is initiated reflects the great latent heat of fusion/crystallization of ice (340×10^3 Joules kg^{-1}) which has to be extracted from water to change it from liquid to solid. Since living organisms consist largely of water, this slowing can be of great biological significance as will be seen in later chapters.

The description of ice formation given here has been for pure water in fairly large volumes. Biological solutions show different behaviour because the diffusional properties of water molecules are altered by the presence of other molecules amongst them. Water in pores (e.g. in protein molecules) also behaves differently from water in bulk solutions. If pores are very small ($<1\,\mu m$), the water within them may be unable to assume the lattice form of ice, and hence be unfreezable (Franks, 1985). This may be an explanation for the so-called bound water often referred to in cryobiological literature.

The most abundant water on earth is in the form of sea water; this has a freezing point of about $-1.9°C$ because of its salt content (which depresses the freezing point). When sea water freezes it rarely freezes whole (i.e. complete with salt content). Usually the fresh water within it tends to freeze out first, so that the remaining water becomes more saline, with a lower freezing point. In high latitude areas this can result in very high salinities ($< 70‰$) beneath ice in small bodies of water such as rock pools. Sea ice can often contain liquid inclusions of equally high salinity.

1.3 COLLIGATIVE PROPERTIES OF AQUEOUS SOLUTIONS

The thermal behaviour of aqueous solutions is affected by the quantity of small molecules in solution. Salinity, body fluid osmolarity and freezing phenomena are inextricably linked, so it is necessary to consider some of the basic concepts and units involved. Sea water in the open ocean has a salinity of about 35‰ (i.e. 35 grams of salts per kilogram of sea water). Sea water behaves osmotically as if it was an approximately molar solution of non-electrolyte, so may be said to have an osmolality of 1 Osmole kg^{-1} (more usually written as 1000 mOsmoles kg^{-1}). Thus a freezing point depression of 1.9°C corresponds to 1000 mOsmoles and there is a nearly linear relationship between osmolality and freezing point depression (particularly in dilute salt solutions), so that 50% sea water (17.5‰) has an osmolality of about 500 mOsmoles kg^{-1} and a freezing point of about −0.95°C. Marine invertebrates, almost all of ancient marine ancestry, have body fluid concentrations very similar to seawater (i.e. about 1000 mOsmoles kg^{-1}) so they become vulnerable to freezing only when sea water (17.5‰) has an osmolality of about 500 mOsmoles kg^{-1} and a freezing point of about −0.95°C. Marine invertebrates, almost all of ancient much more dilute body fluids (broadly 300–500 mOsmoles kg^{-1}) so can freeze before sea water reaches its freezing point.

1.4 CATEGORIES OF BODY TEMPERATURE CONTROL

There are two broad categories of response to temperature; ectothermy and endothermy. Ectotherms are animals in which internal sources of heat production make negligible contributions to body temperature and in which physiological mechanisms for heat retention are weak or absent. Endotherms are animals which have significant internal sources of heat (i.e. from metabolic processes) and have anatomical and physiological mechanisms to control the loss of that heat so that a relatively constant core temperature (i.e. the temperature of crucial organs such as brain, heart, liver) is maintained. In the early days of physiology there was assumed to be a hard and fast division between ectotherms and endotherms; all invertebrates and lower vertebrates (fish, amphibia and reptiles) were believed to be ectotherms; all birds and mammals appeared to be endothermic. With further study this distinction has become blurred. Muscular thermogenesis (i.e. heat production) in flying insects causes their body temperature to be elevated above ambient levels and they can therefore be described as 'exercise endotherms' (they become ectothermic at rest). The best studied insects in this respect are honey bees (*Apis mellifera*). Schmaranzer and Stabentheiner (1988) have recently employed tele-thermography (a technique in which infra-red radiation is recorded from objects to assess their surface temperatures) to study bees. They found that the surface temperature of the bee head, thorax and abdomen can be considerably elevated after a flight. Values of 36, 40 and 31°C were reported from these structures (respectively) in a bee exposed to an ambient temperature of 27°C. The high thoracic temperatures are undoubtedly due to the muscular thermogenesis of the flight muscles. Similar thermogenesis (coupled with efficient vascular counter-current heat exchange systems [q.v.]) permits the swimming musculature of tunny fish to operate at elevated temperature (Carey and Teal, 1966) — a case of localized endothermy. Counter-current heat exchange systems in the flippers, together with peripheral blubber also allow the giant leatherback turtle *Dermochelys coriacea* to maintain a core body temperature heightened by as much as 18 deg. C (Frair *et al.*, 1972; Davenport *et al.*, 1990b) and thereby penetrate temperate waters as far north as Iceland and northern Japan in pursuit of its jellyfish prey.

Conversely, a number of mammals and birds demonstrate what was once known as 'imperfect endothermy', but this has overtones of maladaptation and the term in general use at present is heterothermy. Hibernating mammals (especially bats) show much reduced body temperatures in winter, while humming birds (close to the lower size limit for endothermy anyway) tend towards ectothermy at night when foraging for fuel (nectar) is impossible.

In some older texts 'poikilotherm' replaces 'ectotherm' and 'homoiotherm' replaces 'endotherm'. These older terms are unsatisfactory, since poikilothermy is associated with the concept of a variable body temperature, which some ectotherms, particularly those living at the constant low temperatures of the abyssal depths of the ocean, do not feature. Homoiothermy, on the other hand, carries the implication of constant body temperature (which numerous endotherms do not maintain). Neither of these terms, nor the earlier 'cold-blooded' and 'warm blooded', will be much used in this book. However, one concept, that of 'inertial homoiothermy' needs to be considered. Large ectothermic animals (e.g. adult crocodiles) are so big that they take a great deal of time to equilibrate thermally with their surroundings (by virtue of their low surface-area to volume ratios). Because of this, they are able to maintain fairly constant body temperatures throughout the day and night, without possessing a high metabolic rate.

1.5 ZONE OF NEUTRALITY, CRITICAL TEMPERATURES

For classic endotherms (those animals which maintain an almost constant core body temperature) there is a temperature range over which metabolic rate is constant, the 'zone of neutrality' or 'thermoneutral zone' (TNZ). Below a certain temperature, the lower critical temperature (LCT), the animal has to produce extra metabolic heat to maintain body temperature and therefore

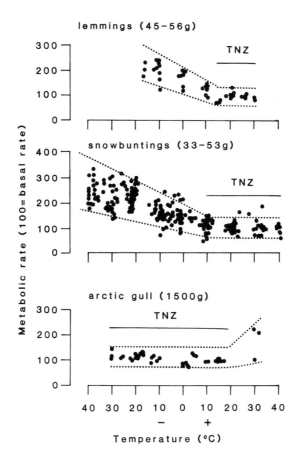

Figure 1.3 Effect of exposure to low temperature on metabolic rate in three arctic endotherms. Redrawn from Scholander *et al.* (1950c). TNZ = thermoneutral zone, the temperature range over which metabolic rate is minimal and unaffected by external temperature.

there is an increase in oxygen uptake. If the temperature continues to fall, the oxygen uptake continues to rise, and it is common to estimate LCT by plotting oxygen uptake against temperature in the manner shown in Figure 1.3. There is also a higher critical temperature (HCT) at the upper limit of the zone of neutrality. Above the HCT the animal has to expend energy in avoiding a rise in body temperature.

The concept of LCT was introduced by Scholander *et al.* (1950b) who measured metabolic rates in a range of tropical and arctic mammals, recording LCT values as low as −40°C for arctic foxes (*Alopex lagopus*). However, such extreme values have recently been criticized by Korhonen *et al.* (1985) who relied upon observations of heart rate and shivering in arctic foxes (as well as measurement of metabolic rate) to derive a LCT of −6°C; it appears that the wild animals used by Scholander *et al.* may have been disturbed throughout experimentation, so these classical results, widely cited for the past four decades, should be treated with some caution.

1.6 NEWTON'S LAW OF COOLING

Sir Isaac Newton was the first scientist to investigate heat losses by warm bodies in air. He found that the rate of loss of heat is proportional to the excess temperature over the surroundings. For inanimate objects this Law holds for excess temperatures of around 20–30 deg. C in still air. If a warm ectotherm is placed in a cool environment and its core body temperature monitored as it cools, a departure from Newton's Law can indicate a degree of physiological control of heat loss.

1.7 THERMAL CONDUCTANCE

Thermal conductance is a measure of the ease with which heat enters or leaves a body. There are four possible pathways of heat exchange between a body and its environment; radiation, conduction, convection and evaporation.

McNab (1980) shows that the thermal exchange of an endothermic animal maintaining a constant body temperature may be described by the following equation

$$Q = \varepsilon \, \sigma \, A_1 \, (T_b - T_a) + k \, A_2 \, (T_b - T_a) + h_c \, A_3 \, (T_b - T_a) + LE \qquad [1]$$

Where Q = rate of metabolism, ε = emissivity, σ = Stefan Boltzman constant, A_1 = surface area for radiative exchange, T_b = body temperature, T_a = ambient temperature, k = thermal conductivity of integument, A_2 = surface area for conductive exchange, h_c = convective coefficient, A_3 = surface area for convective exchange, L = latent heat of vaporization, E = amount of water evaporated.

Clearly thermal exchange is a complex process, and there are several problems involved with any attempt to use equation [1] rigorously. For example; what temperature should be used for T_b? Strictly it should be skin, fur or feather temperature, but in practice these are difficult to measure and core temperature is the parameter of crucial importance to an endotherm. Also, the areas available for radiative, conductive and convective exchange are extremely difficult to measure (and may be controllable by the animal anyway, for example by postural changes).

Consequently there is a need for a simpler relationship, particularly for comparative studies, and the concept of thermal conductance fulfils this role

$$Q = C' (T_b - T_a) + LE \qquad [2]$$

Where C' is the thermal conductance of the animal concerned.

There are still problems with the estimates of thermal conductance. All endotherms have a zone of neutrality where Q is unaffected by temperature. At high temperatures within that zone $T_b - T_a$ will be small, so C' will be high. At the lower limit of the zone of neutrality $T_b - T_a$ will be large so C' will be small — the minimal thermal conductance. Minimal thermal conductances are useful for comparative purposes.

Evaporative water loss is difficult to quantify and, as evaporation tends to be a relatively unimportant route of heat loss at low temperatures (5–15%). Accordingly an even simpler relationship may be used

$$Q = C (T_b - T_a) \qquad [3]$$

Where C = wet minimal thermal conductance, and $C = C' + LE/(T_b - T_a)$.

A commonly written expression of wet minimal thermal conductance is as follows

$$C = VO_2 / \Delta T \qquad [4]$$

Where VO_2 = oxygen consumption rate, and $\Delta T = T_b - T_a$.

If VO_2 is measured as ml $O_2\, g^{-1}\, h^{-1}$ and ΔT as deg. C, then C is measured as ml $O_2\, g^{-1}\, h^{-1}\, deg.\, C^{-1}$.

A detailed critique of graphical and calculative methods for estimating C from VO_2 data is given by McNab (1980). Briefly, he advocates the calculation of a mean conductance value from each of a number of VO_2 measurements made at a range of temperatures below the zone of thermoneutrality.

1.8 WINDCHILL

The air in contact with the skin of an endotherm (or any other hot object, including a sun-warmed ectotherm) forms a thin layer (a few mm thick) which is stationary. Outside this stationary layer the air is moved first by convective currents (warm air rises by virtue of its lower density than cold air), then by movements of the animal itself, and finally by the presence of wind. Heat has to pass through the stationary layer before being lost to the environment. Flowing air makes the stationary layer thinner (effectively stripping much of it away) and heat loss more rapid, so that, for a given temperature, more heat is lost in windy conditions than in still air — the windchill phenomenon. Several

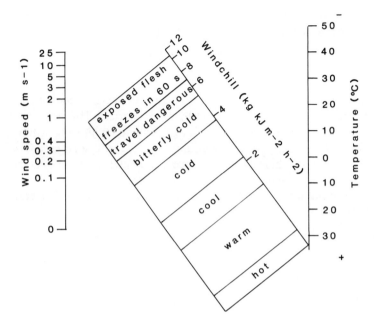

Figure 1.4 Windchill scale. Devised by Siple and Passel from their work in the Antarctic; modified and redrawn from Edholm (1978).

windchill scales have been produced; one is shown in Figure 1.4. From this normograph it can be seen that windchill is not linearly related to wind speed. Relatively modest winds cause greatly increased heat loss above the rates observed in still air, higher windspeeds do not cause proportionally greater heat loss rates. The effect of windchill is considerable; at $-20°C$ an endotherm loses heat about 3 times as fast when exposed to a 90 kph wind as when in still air.

No simple scale is available to describe the interaction between low temperature, wetness and wind. A wet body in an atmosphere of less than 100% relative humidity loses water by evaporation. The latent heat of evaporation (2.26×10^6 J kg^{-1} water) is drawn from the body. Water vapour has to pass through the stationary layer described above, so again flowing air will increase the rate of evaporation (and hence heat loss). If the water on the skin is constantly renewed (either by rain, snow or the sweat of exertion), then the heat required for evaporation must be added to straightforward windchill.

1.9 Q_{10} RELATIONSHIP

Van't Hoff (1884) observed that the rate of chemical reactions tends to double for each 10 deg. C increase in temperature. The rates of biochemical reactions are equally affected by temperature, and biologists and medical researchers have used widely the following Q_{10} relationship to describe the effects of temperature on metabolic processes

$$Q_{10} = \frac{R_1}{R_2} [10/(T_1 - T_2)]$$

Where R_1 = metabolic rate at temperature T_1, and R_2 = metabolic rate at temperature T_2.

The Q_{10} value obtained is therefore the ratio between metabolic rates recorded at two temperatures, but extrapolated to a 10 deg. C difference. Thus, if an ectothermic animal uses 10 ml O_2 at 5°C and 14 ml O_2 at 8°C (actual ratio between rates = 1.4), the Q_{10} for oxygen consumption over the range 5–8°C is calculated as follows

$$Q_{10} = \frac{14}{10} [10/(8 - 5)] = 3.07$$

If an animal exhibits a constant Q_{10} for metabolic rate over a wide temperature range, the metabolic rate is not linearly related to temperature (a common student error!) (Figure 1.5). It should be noted that the Q_{10} relationship has been used for parameters beyond simple oxygen uptake; Q_{10} values for heart rate, ventilation rate, nitrogen excretion and even swimming speed have all appeared in numerous scientific papers.

   The advantage of use of the Q_{10} relationship is that it standardizes the effect of temperature. However, uncritical use of the relationship has many pitfalls. From Van't Hoff's observation one would expect the Q_{10} of biochemical reactions to be close to 2. Physiologists and ecologists have often assumed in energetics calculations that this is true for metabolism itself (and for the crudest of calculations this rule of thumb may be useful), but as long ago as 1916, the great Danish physiologist Auguste Krogh showed that Q_{10} values were not uniform throughout the temperature range (Table 1.1), and his results suggested that Q_{10} rose as temperature fell. A living animal is several orders of magnitude more complex than a simple chemical reaction, and has large numbers of enzyme systems whose characteristics may be changed in response to intracellular or extracellular control mechanisms. In addition, seasonal acclimatory effects, and seasonal changes in biochemistry associated with reproduction are superimposed upon this basic complexity. Finally, poor experimental design may produce anomalous Q_{10} values. Transferring animals

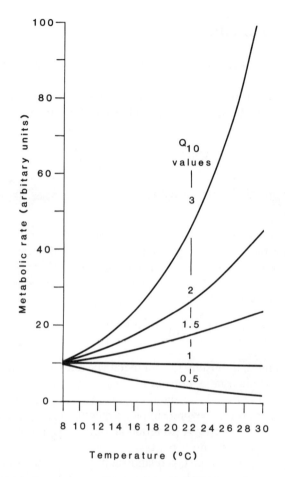

Figure 1.5 Idealized metabolic rates for animals exhibiting steady Q_{10} values (from 0.5 to 3.0) over a wide temperature range. Note that a steady Q_{10} does not imply a linear relationship between temperature and metabolic rate.

from a 'normal' temperature (i.e. one which the animal is likely to encounter in nature) to an abnormally stressful one is likely to produce either a very high Q_{10}, or, if the animal is close to death, a very low value. Special problems are created in those animals which spend part of their existence in a dormant state (usually at low temperature). Q_{10} values as high as 100 have appeared in the literature as a result of comparing the metabolic rate of the dormant stages with that of active animals. The author has some sympathy for the views of Johnson *et al.* (1974) who stated that there is no rational basis for the application of the Q_{10} relationship to whole animals!

Table 1.1 Variation of Q_{10} with temperature

Temperature range (°C)	Q_{10} for fish amphibia and curarized dog*	Q_{10} for Tenebrio pupae
0–5	10.9	–
5–10	3.5	–
10–15	2.9	5.7
15–20	2.5	3.3
20–25	2.2	2.6
25–27.5	2.2	2.3
27.5–30	–	2.1
30–32.5	–	2.0

* use of a curarized dog allowed Krogh artificially to alter the body temperature of an endotherm. After Krogh, 1916.

1.10 COLD ACCLIMATION AND COLD ADAPTATION

1.10.1 Definitions

Acclimation and adaptation are widely used terms in thermal biology, yet are still often confused. Acclimation to cold takes place when an animal is exposed to a colder environment than it was living in previously, and undergoes changes to its anatomy, physiology, biochemistry (and conceivably behaviour) which make it perform better in the cold environment. Acclimation is phenotypic, needing no alteration in genetic material; it operates at the individual level. Under natural conditions acclimation is largely a seasonal phenomenon and tends to involve rather slow changes in average temperature. In the laboratory it has been quite common for animals to be abruptly moved from one thermal environment to another — this approach risks the collection of unphysiological results (see Davenport, 1982 for discussion). As a general rule complete acclimation takes place over a period of weeks or months.

Adaptation to cold operates at the species level and involves changes in genetic material. Adaptation to cold takes place when species are either exposed to cooling conditions, or expand into cold areas. As with acclimation, changes in anatomy, physiology, biochemistry and behaviour may all be necessary for a species to cope with a cold environment (throughout the seasons). Adaptation takes place in response to natural selection; acclimation forms a subset of the adapted animal's capabilities.

1.10.2 Cold acclimation

Acclimation to cold tends to mean different things in ectotherms and endotherms, basically because the tissue temperature of the former changes when they are exposed to cold, while in the latter, the temperature of 'core' tissues does not change.

In ectotherms acclimation can be considered under two headings, metabolic acclimation and ultrastructural acclimation, although there is overlap at the molecular level. Metabolic acclimation was categorized by Precht (1958). An ectothermic animal would be expected to show a lowered metabolic rate at a lower temperature, simply because of the Q_{10} effect described above. Precht described various sorts of compensatory response (Figure 1.6).

Precht type 1 response   Overcompensation

An animal showing a Precht type 1 response to cold will show a higher metabolic rate when it is acclimated to a low temperature than it did when previously held at a higher temperature. This type of response is rare and has the obvious disadvantage that extra energy is consumed at lower temperatures. (Q_{10} over temperature range of acclimation less than 1).

Precht type 2 response   Perfect compensation

In this response, the animal concerned will usually show a reduction in metabolic rate when it is moved from a warm to a cooler environment, but, when it has fully acclimated, the metabolic rate returns to the original level. A number of bivalve molluscs (e.g. the mussel, *Mytilus edulis*) show perfect compensation over limited temperature ranges. The type 2 response has the advantage that the animal concerned maintains a constant rate of living despite

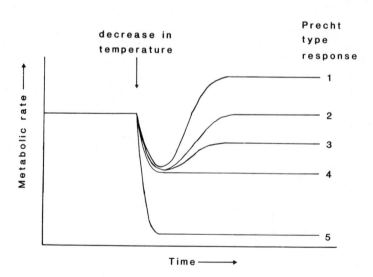

Figure 1.6 Graphical representation of various types of acclimatory metabolic response to a decrease in environmental temperature. See text for details.

changes in the thermal environment. Other terms for a type 2 response are metabolic independence or homeokinesis — the maintenance of kinetic energy within a system (Q_{10} close to 1).

Precht type 3 response   Partial compensation
In the type 3 response, the acclimated animal shows a higher metabolic rate than it would if simple Q_{10} effects were operating, but a lower metabolic rate than the one operating at the original higher temperature. Partial compensation is by far the most common response of invertebrate and vertebrate ectotherms and is particularly well documented for active polar mites and insects. (Q_{10} greater than 1, but less than 2).

Precht type 4 response   No compensation
The situation when an animal shows no acclimatory or compensatory response to a temperature change; the animal's oxygen uptake or other rate functions follow a simple Q_{10} relationship (Q_{10} close to 2).

Precht type 5 response   Paradoxical or reverse acclimation
Shown in animals which respond to cold by dormancy or torpor. The metabolic rate of animals moved and acclimated to a cold environment is lower than would be expected from simple Q_{10} considerations. This approach is adaptive in situations where cold is associated with shortage or absence of food. It is seen in amphibia, reptiles and invertebrates such as insects and some species of land snails (e.g. Bailey and Lazaridou-Dimitriadou, in press). (Q_{10} values above 2, often at very high levels in arthropods from high latitudes.).

Precht's classification of metabolic acclimation is still useful, but, as so often in the untidy science of biology, there is much gradation of response. Mussels, for example, show more or less complete compensation (type 2) between 10°C and 25°C, but show partial compensation (type 3) at lower temperatures (Bayne, 1976). There is even evidence of type 5 response in mussels with frozen body water at subzero temperatures!

In those ectotherms which show Precht type 2 or 3 acclimatory responses, there have to be changes in the tissues which allow raised metabolic rates as acclimation progresses. Dunn (1988) reports that a wide range of teleost fish show increased mitochondrial numbers in the muscles when cold acclimated, indicating increased oxidative capacity. Kent *et al.* (1988) have recently studied the freshwater channel catfish (*Ictalurus punctatus*) which shows partial compensation (type 3 acclimation) over temperatures from near 0°C to 30°C. Catfish were moved from a temperature of 25°C to water at 15°C and allowed to acclimate for 4 weeks. The fish showed profound anatomical, histological and biochemical changes to the liver. Liver mass, cell size, liver protein content and total liver enzyme activity all roughly doubled. The enlarged cells also contained about 4 times as much glycogen as did fish at 25°C. Specific enzyme activities were lower at 15°C than at 25°C (as would be expected from simple Q_{10} considerations), but much of the increased protein content was due to increases in enzyme content, so that rates of metabolic activity could be sustained. Kent *et al.* also found hypertrophy of the heart of catfish, though this was less dramatic than the hypertrophy of the liver. They also surveyed the acclimation literature and showed that changes in mass of the liver were common in cold acclimated teleosts; it is probable that similar combinations of structural and metabolic response operate in cold-acclimated invertebrates too.

Studies of ultrastructural/molecular acclimation have tended to concentrate on the structure of cell membranes. For nearly twenty years, biologists have recognized that cell membranes are composed of lipid bilayers with associated proteins (extrinsic proteins associated with the hydrophilic polar heads of the lipids, intrinsic proteins within the hydrophobic hydrocarbon domain of the lipid layers). This is the fluid mosaic model of Singer and Nicolson (1972) (Figure 1.7). However, as study has progressed, the nature of the lipid layers has been recognized as extremely complex (see Shinitsky, 1984a, b for comprehensive reviews). Quinn (1985) points out that even simple animals such as ciliate protozoans may have more than 40 different lipid molecular species in the membranes (classified by fatty acid chain length, degree of unsaturation and site of attachment of acyl chains to the glycerol moiety). Lipid membranes contain water and the form that they adopt is influenced by the amount of water in the membrane (in turn influenced by factors such as salt concentration) and by temperature. As many as 11 phases of lipids are

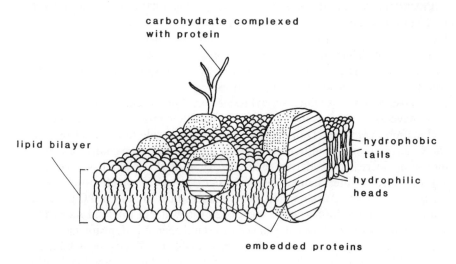

Figure 1.7 Fluid mosaic model of membrane structure.

described by van der Meer (1984) and phase changes (e.g. those induced by cooling) result in different forms of membrane 'order' or 'fluidity'. Broadly speaking, lowered temperatures cause decreased membrane fluidity and this, in turn will affect the ability of the membrane to allow the passage of material through it, and will also upset the balance between passive fluxes and active pumping of salts. Damage by cold to membranes is caused by more extreme phase changes which do not reverse on rewarming.

Cold-acclimated ectotherms demonstrate 'homeoviscous responses' or 'conservation of membrane fluidity' (Cossins and Sinensky, 1984). This is achieved by changes in the composition of membrane lipids to offset the effects of the reduced kinetic energy of low temperature. The most conspicuous change in composition is an increase in the proportion of unsaturated fatty acids in the hydrocarbon domain of the membranes. Hazel and Sellner (1980) point out that unsaturated fatty acids have a lower melting point than their saturated homologues; they also occupy greater areas of membrane film and are individually more fluid. Table 1.2 shows the changes in fatty acid composition induced by cold acclimation in rainbow trout (*Salmo gairdneri*). It is evident from this table that the changes in composition are not simply a matter of substituting unsaturated for saturated fatty acids. In the trout liver cells there is a tendency (within the pool of unsaturated fatty acids), for the accumulation of unsaturated fatty acids with longer chains. Hazel and Sellner provide extensive evidence of seasonal, cold-induced restructuring of membranes.

Table 1.2 Effects of cold acclimation on fatty acid composition (presented as % by weight) of hepatocytes on rainbow trout

Fatty acid class	Cold-acclimated trout (5°C)	Warm-acclimated trout (20°C)
Saturated fatty acids	25.8	28.7
Unsaturated fatty acids		
Total	75.2	71.1
monoenes	23.9	36.2*
dienes	8.5	7.8
trienes	2.6	2.1
tetra/pentenes	11.6	6.7*
22:6n3	24.3	20.8
Polyunsaturates (:4–6)	35.9	27.5*
n–3 family	34.5	27.0*
n–6 family	15.4	14.6
n–9 family	21.9	24.3
Unsaturation index	251.5*	201.3*

* indicates significant difference between warm- and cold-acclimated fish.
Unsaturation index is the summed product of weight percentage and the number of double bonds.
Simplified from Hazel and Sellner, 1980.

They believe that this is achieved by regulation of the composition of the fatty-acyl coenzyme A pool from which membrane lipids are synthesized, or by regulation of the uptake of fatty acids from the pool. A complicating factor is that the fatty-acyl coenzyme pool can be supplied with fatty acids from the diet, or synthesized *de novo*. If supply of fatty acids is mainly from the diet and particular fatty acids can move through food chains, as indicated by Sargent *et al.* (1987), then some of the homeoviscous response in predators may be passively provided by their prey.

Seasonal acclimation in endothermic animals tends to be controlled by photoperiod rather than temperature (e.g. Heldmaier and Steinlechner, 1988), so 'cold acclimation' is often a misnomer in this group. Many different strategies are employed by mammals and birds and they are complicated by food availability, seasonality of reproduction, and whether hibernation is employed or not.

1.10.3 Cold adaptation

Most of this book is concerned with the adaptations of animals to cold; all that will be discussed here are the concepts of metabolic and ultrastructural adaptation, which in functional respects have similarities to the cold acclimatory responses described above, although the timescale of adaptation is clearly very much greater than the timescale of acclimation.

Much attention has been paid to the concept of metabolic adaptation to cold in ectotherms, which seems first to have been considered by Auguste Krogh (1916). He stated, 'One would expect that animals living at very low temperatures should show a relatively high standard metabolism at that temperature, compared with others living normally at high temperature'. This statement is rather ambiguous, but for many years it was generally accepted that high latitude aquatic ectotherms (fish and invertebrates) had a higher metabolic rate at low temperature than would be predicted by extrapolation from measurements of metabolic rate in their temperate/tropical relatives within the normal temperature ranges experienced by those relatives (i.e. by using Q_{10} relationships), and certainly higher than the rates exhibited by temperate animals transferred to polar thermal conditions. This relatively high metabolic rate (raised as much as 10-fold according to Scholander *et al.* (1953) and Wohlschlag (1964)) would indicate a degree of metabolic compensation or adaptation, and would require evolution of suitable enzyme systems and enzyme concentrations to sustain it. Much of the early work involved comparisons between arctic/antarctic aquatic animals and temperate species exposed to around 0°C. This was in many ways unsatisfactory as the temperate animals were often moribund and showed low metabolic rates and very high Q_{10} values at low temperatures in consequence. Later work has tended to involve statistical extrapolation from data collected in the normal temperature ranges of the animals being compared, and is open to the criticism that such extrapolation is unjustified. Clearly it is difficult if not impossible to devise experimental approaches which allow unequivocal interspecific comparisons. It is also worth noting that comparisons involving terrestrial species are even more difficult as the environmental temperatures of ectotherms fall well below 0°C at high latitudes, and so metabolic responses are often complicated by the presence of cryoprotectants and/or ice in the body fluids.

Over the past decade metabolic adaptation has been revealed as a far less clearcut phenomenon than once thought. Clarke (1980, 1983) has argued that faulty respirometric technique and experimental design have often over-estimated the metabolic rate of polar fish and marine invertebrates. For example, he calculates that arctic char (*Salvelinus alpinus*) use no more than twice as much oxygen as acclimated goldfish (*Carassius auratus*). In similar vein, Houlihan and Allan (1982) demonstrated that the oxygen consumption of a variety of Antarctic gastropods are close to the predicted values derived from temperate gastropods, thus arguing against metabolic adaptation. Hochachka (1988) and Dunn (1988) have both pointed out that the life style of animals must also be considered. Although Antarctic fish living in productive inshore waters do tend to have quite high metabolic rates, well within the ranges exhibited by temperate species, bathypelagic teleosts, living at equally low temperatures but at great depth, have far lower metabolic rates (e.g. Childress and Somero, 1979), by as much as 1–2 orders of magnitude, presumably because they are adapted to an environment in which food supply is low and unpredictable. Hochachka (1988) has turned Krogh's statement on its head and asked why an Antarctic fish living at low temperature should use energy (and turnover ATP) at a faster rate than a related temperate species would if it could survive cooling to the same temperature. An ecologist would probably answer that too low a metabolic rate would restrict scope for activity to an unrealistic extent in an energy-rich environment (remember that Q_{10} effects are exponential, not linear). However, Hochachka believes that Antarctic and abyssal fish have had to respond over an evolutionary timescale to a potential osmotic/ionic problem. While passive ion fluxes (physical processes) are little affected by temperature ($Q_{10} = 1.2-1.4$ according to Hille, 1984), ion pumps (ATP dependent) are affected in their efficiency ($Q_{10} = 2-4$). Thus, as temperature falls, there will be a tendency for a mismatch between passive fluxes and active pumping. This is soluble in two ways: by an increase in the number of pumps (pumps being ion-sensitive ATP-ase molecules sited on cell membranes), or by a reduction in the number of channels in the membranes which allow passive ion flux, thus allowing a reduced pump number. Hochachka postulates that the Antarctic fish have evolved the meta-bolically expensive route of increased pump densities and so have a high metabolic rate, while deep water fish have evolved reduced permeabilities, so can maintain the very low metabolic rate appropriate to their cold and energy-poor environment. This is a seductive hypothesis, but there are some problems with it. Hochachka claims that Antarctic fish have 'standard osmotic capacities'. This is debateable since most Antarctic fish have blood osmolarities of around 600 mOsm kg^{-1} — nearly double the blood concentration of cod, capelin and plaice living in the Arctic basin. He also implies that bathypelagic fish have poorer osmotic control than Antarctic fish. The evidence for this is rather limited as physiological studies on live bathypelagic fish have been few indeed (and necessarily carried out at pressures very different from ambient, which are likely to have profound effects on membrane function). However, there is no doubt that the pump/channel relationship is an important one, which must be taken into consideration. Hochachka also makes one telling point that must not be forgotten: 'Antarctic fishes are demonstrably metabolically cold adapted (they could not metabolize anything at any rate at all [at $-2°C$] if they were constructed of the same macromolecules as are found in rats!)'.

Another piece of evidence in favour of metabolic adaptation at high latitudes is that recent measurements of oxygen uptake and other rate functions of polar marine animals have revealed low Q_{10} values as long as measurements have been made within the narrow limits of temperature which the animals encounter in nature (rather than at the unphysiological and stressful temperature ranges employed in the past). For example: the author (Davenport, 1988) investigated oxygen consumption and ventilatory frequency in the common Antarctic protobranch bivalve mollusc, *Yoldia eightsi*. Oxygen uptake had a Q_{10} of about 1 between 0.2 and 2.5°C; ventilatory frequency changed little ($Q_{10} = 1.72$) over the range 0–3°C. Having a metabolic rate which is relatively insensitive to temperature will have ecological advantages for the species, since it will minimize loss of energy reserves during periods of limited food supply.

Cold adaptation at the ultrastructural level has been studied in terms of membrane lipids (which tend towards greater chain length and degree of unsaturation of fatty acids in high latitude animals), and in the structure of muscles. Johnston's group at the University of St Andrews have carried out a great deal of work on the muscles of nototheniid fish of the Antarctic in the past decade. Much of their work is outside the scope of this volume, but Johnston and Altringham (1988), used de-membranated fast ('white') muscle fibres to study the thermal properties of muscle proteins. They showed that the muscle proteins were only stable at low temperature. They also found that the maximum tension developed was relatively insensitive thermally within the normal environmental range ($Q_{10} = 1.1–1.3$), and was greater than that recorded in temperate and tropical fish. Contraction velocities and power outputs of muscles are less than those shown by warm-water fish at their preferred environmental temperature, but substantially greater (roughly 10-fold) than those exhibited by warm-water fish cooled to 0°C. It appears that the muscles of Antarctic fish (adapted to cold over a much longer period of time than arctic fish) are structurally adapted, but the adaptation cannot completely compensate for low temperature — which probably has a limiting influence on locomotory performance.

1.11 HEAT PRODUCTION

1.11.1 General considerations

Endotherms, like ectotherms, gain heat from the environment, in the form of relatively short wave length radiation from the sun (0.3–3 μm) and rather longer wavelengths (3–100 μm) from warm surroundings (rocks, trees, earth, sand etc.). They can also modify their heat balance by basking in the sun or shade-seeking. However, their main source of the heat necessary to maintain a differential between a high core temperature and a cold environment lies in metabolic heat. All oxidative breakdown of biochemical substrates yields heat, and there is an inevitable production of heat from all metabolic processes, many of which are no more than 20% efficient. This is true of ectotherms' metabolism too; endothermy stems from the production of larger quantities of heat beyond this basic level. In birds and mammals heat is apparently generated mainly in the liver and muscles which can make up more than half of the total body weight, though 'emergency' heat may also be produced by other tissues, particularly brown adipose deposits [q.v.]. Clearly endothermy is more expensive in energetic terms than ectothermy, but by how much? Physiologists have been interested for many years in the relationship between metabolic rate and body size in a wide variety of ectotherms and endotherms. The relationship is expressed in the following exponential formula

$$Q = k\,W^b$$

where Q = metabolic rate (usually measured as oxygen consumption), W = body weight, k and b are constants.

In calculations it is often more convenient to use the following transformed relationship

$$\log Q = \log k + b \log W$$

If $\log Q$ is plotted against $\log W$, a straight line of slope b is obtained. If metabolic rate is simply proportional to body weight then $b = 1$. This rarely occurs save in some colonial organisms (e.g. bryozoa), echinoderms and coelenterates, where growth tends to be more nearly two dimensional than three dimensional. Early workers suspected that b should be close to 0.67, indicating that metabolism is proportional to surface area. Values from about 0.45 to 1.0 have been recorded in the literature, but thirty years ago Hemmingsen (1960) collected together a most comprehensive array of data for ectotherms and endotherms which indicated a general b value of 0.75. Hemmingsen standardized all his data for ectotherms at 20°C and his information for endotherms at 39°C (Figure 1.8). It can be seen that the line for endotherms, although having a similar b value to ectotherms, has a much higher k (intercept) value, suggesting that endotherms have a standard metabolic rate some 10–15 times as high as ectotherms of similar size. At high environmental temperatures close to the body temperatures of endotherms (i.e. 35–40°C) it might be expected from crude Q_{10} estimates that endotherms would only use 2–3 times as much oxygen as ectotherms. Conversely, at low temperatures, the tissue respiration of ectotherms would fall while that of endotherms would rise, further increasing the cost of endothermy. Even when assuming a 10–15 fold difference in metabolic rate it would appear that endothermy is very expensive in energetic terms. However, it must be remembered that endothermy allows far higher levels of activity to be sustained for much longer periods and liberates animals from the influences of short term fluctuations in environmental temperature. However, it also means that endotherms have to find and eat far more food. A particularly graphic example of this difference between ectotherms and endotherms was given by Cott (1961) who demonstrated that a Nile crocodile (c. 45 kg) ate far less (absolutely not relatively) than did a large bird such as a

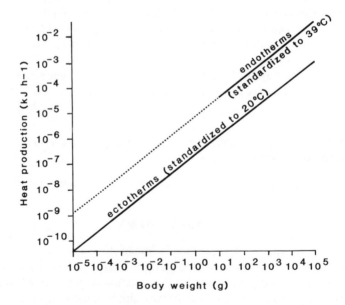

Figure 1.8 Relationship between metabolic rate and body weight in ectotherms and endotherms (solid lines). Dotted line is projected from endotherm data to demonstrate an intercept value more than an order of magnitude higher than for ectotherms. Recalculated and redrawn from Krogh (1916) and Hemmingsen (1960).

pelican (c. 5 kg). Lest this might be thought a strange example, it should be remembered that crocodiles are the closest living relatives of birds, some taxonomists classifying both groups as archosaurs.

The cost of endothermy, combined with the metabolic rate/weight relationship described above, sets a lower limit to the size of endotherms. Because small endotherms use relatively more oxygen, weight for weight, than large endotherms, the relative cost of endothermy is also high (especially because a high surface-area to volume ratio promotes heat loss). Consequently a point is reached where the time needed to find enough food to support a high metabolic rate is such that insufficient time is left for other activities. In both birds and mammals the absolute minimum size limit for an adult animal is about 3–4 g. Shrews of this size must eat very frequently; they starve to death in as little as 4 hours if deprived of food (hence Longworth traps designed to catch them must never be left without food within them and must be inspected often). The smallest birds are humming birds of similar weight. They eat nectar and cannot forage at night, partly because they find flowers visually, but also because the flowers close up at night and the nectar becomes unavailable. Consequently, the smaller humming birds become torpid and near-ectothermic at night to avoid starvation before dawn. Whilst considering the lower size limit for endotherms it should be noted that the young of many endotherms are smaller than 3–4 g (this is true not only of relatively small placental species such as mice, but also the young of fairly large marsupials, because intrauterine development in marsupials is extremely short). These young are effectively ectotherms and need bodily contact with the mother, warm nests or the presence of siblings huddled around them if they are to survive. It should also be noted that the very smallest endotherms (shrews, humming birds, honey possums) have specialized high energy diets (insects, nectar). Animals relying on lower energy diets, particularly herbivores, cannot be quite as small. Hochachka *et al.* (1988) have recently suggested that there may be molecular and ultrastructural limits to body size in endotherms as well as the ecological ones described above. There are limits to the degree to which catalytic efficiency can be improved, so most of the increased aerobic capacity of small endotherms must be provided by increased numbers of mitochondria within each cell. Beyond a certain point, the numbers cannot be increased without compromising other aspects of cell function.

The high energy requirements of endotherms, particularly in cold climates, have a number of knock on effects, some not immediately obvious. At the individual level, endotherms need to have relatively large amounts of heat-producing tissues. Endotherms tend to have larger livers (to provide heat by non-muscular thermogenesis [q.v.]) than do ectotherms of equivalent size. One would also expect endotherms to have relatively more muscle tissue too (to provide heat by muscular contraction — e.g. in shivering). However, analysis here is difficult, since vertebrate muscle can act either aerobically for long periods ('red muscle') or anaerobically for short periods ('white muscle'). An ectotherm such as a fish has a great deal of muscle, but most is white muscle, only used during short bursts of violent activity. In contrast, endotherms have a higher proportion of red muscles.

There are also effects at the ecosystem level; the large amount of food needed to fuel endothermy means that endotherms cannot be as numerous as ectotherms, since much plant material is converted into heat rather than flesh. In turn this concept demands that numbers of endotherms per unit area of land should fall with decreasing temperature, since the cost of fuel (or the cost involved in offsetting fuel wastage by the production and carriage of enhanced insulation, i.e. fur, fat or feathers) will rise with decreasing environmental temperature. This last point is difficult to support with hard evidence because of the problems of deciding on comparable ecosystems (not to mention the problems caused by Man's effects on the numbers of herd animals throughout the world!). However, if cold grasslands (e.g. the steppes of Mongolia) are compared with the grassy plains of Africa, it seems that the latter support much larger numbers of endothermic animals.

It has also been suggested that limited food availability sets constraints on the maximum size of endothermic herbivores (with living elephants and extinct mammals such as the 5–6 m high rhinoceros of the Oligocene, *Baluchitherium*, approaching the upper limit — unless sauropod dinosaurs were really endotherms! (Chapter 9)). Although such animals have a relatively low metabolic rate on a weight specific basis, they still require large quantities of food. Much of the plant material around them is unavailable, either because it is too short, too spiny, too poisonous or of inadequate nutritional value (Freeland and Jansen, 1974). Finding sufficient food amongst an assemblage of unsuitable plants is likely to become more difficult as a species grows larger. There have to be trade-offs amongst body size, necessary foraging area per individual and the maintenance of an adequate population density for successful reproduction. This situation probably makes large herbivores particularly vulnerable to climatic change, since shifts in climate invariably result in alterations in vegetation.

Also interesting in an ecological context are ratios between then numbers of predators and prey. Bakker (1975a,b) has contended that predator : prey ratios are much lower for endotherms than for ectotherms because endotherms need to eat much more to sustain their body temperature. He has used this concept to support his hypothesis that dinosaurs were endotherms. While Bakker's 'dinosaur heresies' are controversial, there is general support for the idea that endothermic predators must be few in number. A good account is given by Hotton (1980), who demonstrated that this effect is greatest for small predators, but becomes more difficult to discern in the case of large predators (of the size of pumas and Komodo dragons). In the case of shrews and lizards the effects are dramatic with predator : prey ratios (on a biomass basis) of 0.02 and 0.50

respectively (i.e. shrews being 25 times less numerous than lizards). For pumas and the ectothermic *Varanus komodoensis* the equivalent ratios are 0.13 and 0.25.

1.11.2 Thermogenesis

As indicated above heat is produced by catabolism — the breakdown of biochemical substrates. Heat production has been particularly studied in the breakdown of glucose molecules. In the presence of oxygen, glucose is broken down in the cytoplasm of cells by glycolysis (also known as the Embden-Meyerhof pathway) to pyruvate and thence to acetyl coenzyme A (acetyl CoA). One molecule of glucose yields two molecules of acetyl CoA. During glycolysis there is a net gain of two adenosine triphosphate (ATP) molecules for each molecule of glucose broken down; four hydrogen atoms (protons) are also produced. ATP molecules (produced by the phosphorylation of adenosine diphosphate, ADP) are energy storage molecules, the energy being stored as 'high energy bonds'. Acetyl CoA is then transferred from the cell cytoplasm to mitochondria. Mitochondria are small organelles (1.5–10μm in length; 0.25–1.0μm in diameter) (Figure 1.9) surrounded by a double membrane, the inner one being folded to form cristae. In the central matrix of the mitochondrion the next stage of energy extraction/storage takes place. Acetyl CoA enters the tricarboylic acid cycle (or Krebs' Cycle) by combining with oxaloacetate to form citrate. Oxaloacetate itself is eventually formed from citrate (via a number of intermediates) to complete the cycle. During one turn of the cycle

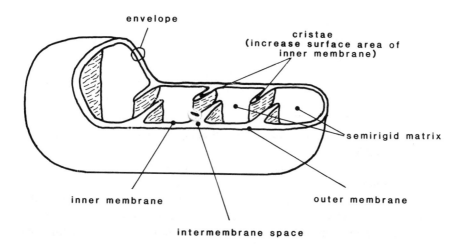

Figure 1.9 Diagram of a mitochondrion. Mitochondria are organelles ranging from 1–10 μm in length.

two molecules of acetyl CoA yield two further molecules of ATP, six molecules of CO_2 and twenty hydrogen atoms. Thus, at the end of the tricarboxylic acid cycle a molecule of glucose has yielded four molecules of ATP and twenty four hydrogen atoms. Twenty hydrogens reduce nicotinamide adenine dinucleotide (NAD) to $NADH_2$, the remaining four hydrogens reduce flavin adenine di-nucleotide (FAD) to $FADH_2$. The hydrogen atoms are then transferred (with the liberation of electrons) from the dinucleotides to the respiratory (= electron transport) chain, the components of which (cytochromes) are located on the cristae of the inner mitochrondrial membrane. The twenty protons derived from $NADH_2$ eventually combine with five molecules of oxygen and twenty electrons to yield ten molecules of water. During their passage along the chain the electrons also yield thirty ATP molecules. The four protons associated with FAD yield four molecules of ATP during passage of electrons along the chain, so in total the ATP yield from a glucose molecule is as follows

From glycolysis	2 ATP
From tricarboxylic acid cycle	2 ATP
From respiratory chain	34 ATP
Total	38 ATP molecules

Now, the energy liberated by complete oxidation of a mole of glucose is 2870 kJ. The energy stored by phosphorylation of a mole of ADP to ATP is about 50 kJ (the precise value depends upon pH and temperature — see Brafield and Llewellyn, 1982 for discussion). So, a yield of 38 ATP moles should store 1900 kJ, leaving 970 kJ to be lost from the cell as heat. During normal aerobic metabolism some 34% of the energy derived from carbohydrate breakdown will therefore be liberated as heat. Similar considerations apply to the break-down of other substrates (e.g. β-oxidation of fatty acids derived from neutral lipid stores, which also yields acetyl CoA to be fed into the tricarboxylic acid cycle). Indeed fatty acids are extremely important sources of energy and heat because of their high catabolic yields of ATP (a mole of stearic acid produces 147 moles of ATP on complete oxidation).

In endotherms more heat is liberated than in ectotherms because tissues are more active. They have more numerous mitochrondria (particularly in the case of muscle and liver) and there is a greater turnover of substrates. However, the biochemistry of heat production by both groups of animals is essentially similar for most tissues. There are exceptions though, and these are related to 'emergency' heat demands. Rapid-flying insects (e.g. honey bees) are part time endotherms; their wing muscles need to be warmed before they function effectively. Many texts imply that muscle temperatures are raised simply by muscular activity before take off. However, it would appear that primary heat to warm the muscles before activity starts is generated by 'futile cycling', the non-productive flux of substrates through a series of anabolic and catabolic

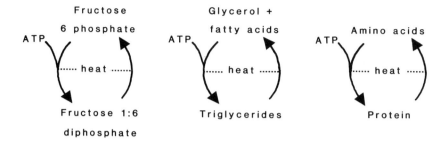

Figure 1.10 Possible futile, heat-generating biochemical cycles.

reactions simply to generate heat (Figure 1.10 for possible cycles). Such futile cycling is feasible in most animals, but its occurrence and importance is generally unclear, with the exception of the processes which take place in brown adipose tissue (BAT).

Brown adipose ('brown fat') has been identified and studied most in mammals, though there are indications that it may be present in other endotherms, including leatherback turtles (Mrosovsky, 1980). Brown fat is most noticeable in very young mammals (where it is important in counter-acting chilling in cold weather), and in adults of hibernating mammals where it is crucial to the rapid warming associated with arousal. However, work on humans has revealed that brown fat persists into adulthood; this may be true of other non-hibernating mammals too. In newborn rabbits brown fat can make up 5% of the body weight, being concentrated in the neck and thoracic regions. Brown adipose gains its name because it is packed with mitochondria containing cytochromes of the respiratory chain; the cytochromes provide the brown colour which contrasts with the white or yellow appearance of normal adipose tissue. Brown fat is invariably well vascularized (in contrast to storage fat deposits which often have scanty blood supplies).

Most tissues which are rich in mitochondria are also laden with ATP. In the case of brown adipose tissue this is not the case, because levels of ATP synthetase are low and the tissue is capable of a specialized breakdown of carbohydrates and fatty acids in which energy is lost as heat rather than stored as ATP. For most of the time brown fat is a fairly low activity tissue in which carbohydrates and fatty acids eventually supply protons to the cytochromes of the respiratory chain located on the inner mitochondrial membrane (Figure 1.11a) as in normal catabolism. The respiratory chain transfers protons from the matrix of the mitochrondrion to the outside, creating a pH gradient across the membrane. Production and maintenance of this gradient is an energy consuming process (the energy being provided by movement of electrons along the respiratory chain), so the gradient (or proton motive force, PMF) effectively functions as a store of potential energy which is drawn on as protons

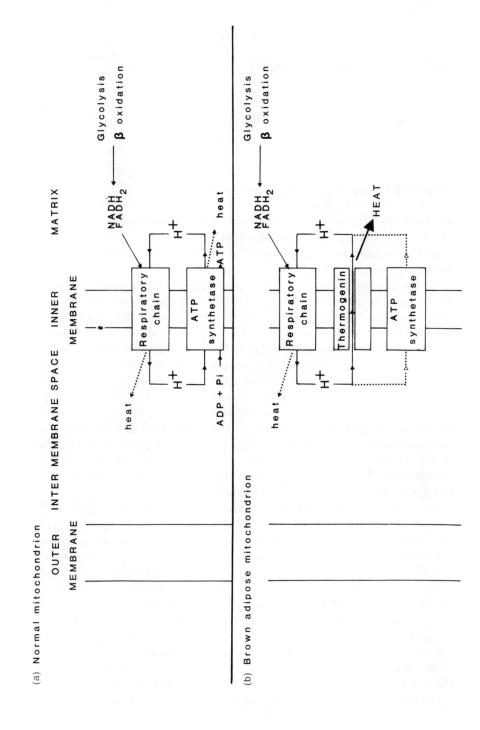

(a) Normal mitochondrion

(b) Brown adipose mitochondrion

move back across the membrane. Most of the mitochrondrial membrane is impermeable to protons, which can only return to the matrix through localized channels. When brown adipose is not generating heat, the protons return to the matrix through channels which are associated with the enzyme protein ATP synthetase, and much of the energy (perhaps two thirds) is stored as ATP, the remainder being lost as heat. Under these circumstances the rate of substrate breakdown is controlled by the availability of ADP from which ATP is synthesized.

Heat output by brown adipose is triggered by noradrenaline released by sympathetic nerve endings. Additional channels through the inner mitochrondrial membrane become available. These channels are provided by a membrane protein with a molecular mass of 32,000, known either as proton translocase or thermogenin. There is no associated ATP synthetase, so energy storage does not occur as the protons stream through the thermogenin channels back into the matrix; instead all energy is liberated as heat (Figure 1.11b). As no energy storage takes place, the feedback relationship between ADP concentration and proton flow rate breaks down, so the breakdown of lipids and carbohydrates to release heat is limited only by the availability of substrates. In consequence the uncontrolled heat output of active brown fat can be many times that of the inactive tissue. Heat produced in brown adipose tissue is quickly carried away to other parts of the body by the rich blood supply and drainage.

Heat production by the activity of brown fat is a major component of 'non-shivering thermogenesis' (NST) a term used to distinguish heat producing biochemical processes from the heat production due to non-locomotory muscular activity ('shivering thermogenesis'). There are other sources of NST such as the futile cycles described in Figure 1.10 above, and the phenomenon of 'specific dynamic action'. Specific dynamic action (SDA) is a term used to describe the increase in metabolic rate seen after a meal. Such an increase is seen in ectotherms and endotherms alike and is reflected in increases in oxygen consumption and heat output. In endotherms the terms 'heat of nutrient

---

Figure 1.11 Comparison of function in the mitochondria of normal cells and in brown adipose tissues. In normal mitochondria protons pumped into the intermembrane space return to the mitochondrial matrix via ATP synthetase and generate ATP in the process. This sequence results in the storage of energy and the production of small quantities of heat. In active brown adipose tissue most protons return to the mitochondrial matrix via thermogenin (which forms a proton conductance pathway through the inner membrane of the mitochondrion) and little or no storage of energy as ATP takes place. Instead, large quantities of heat are produced as long as carbohydrate and fatty acid substrates are available. In resting brown adipose tissue the functioning of thermogenin is inhibited by the presence of purine nucleotides; this inhibition is counteracted by the presence of noradrenaline.

metabolism', 'calorigenic effect of food' and 'post-prandial thermogenesis' have been used as alternatives to SDA. Dieticians tend to use the term 'dietary-induced thermogenesis' (DIT). The cause of SDA/DIT is obscure, but the thermogenic effect can be great; Lavoisier long ago showed that the metabolic rate (and heat output) of a fasting human rose by 50% after a large meal, and remained elevated for some hours. It seems that SDA is in some way necessary to allow the animal concerned to take advantage of the energy and nutrient content of the meal. Amino acid metabolism has often been implicated as a cause of SDA, but even here it is unclear whether heat is being released during the deamination and elimination of amino acids, or during the synthesis of proteins from them! SDA is often portrayed as an unproductive waste of part of a meal's energy, but in cold climates SDA can be crucial to survival. Certainly regular adequate meals which top up SDA throughout the day tend to stave off hypothermia in elderly people during cold winters, and it must be remembered that the endothermy of young mammals and nestling birds is often fragile, only sustainable in cold conditions because of frequent meals, which have thermogenic as well as nutritive qualities.

2 The cold environment

2.1 PRESENT CONDITIONS

2.1.1 Air temperatures

Most of the world is cold by human standards. It is little appreciated that the average global ground-level air temperature is only about 9°C (integrating the atmospheric temperatures of all latitudes and all seasons). Figure 2.1 shows the mean surface air temperatures at various latitudes in midsummer and mid-winter. It is evident that temperatures decline towards the poles, and that the seasonal amplitude of temperature change is greater at higher latitudes. It may also be seen that there are differences between the northern hemisphere and the southern hemisphere, with the south being substantially colder at equivalent latitudes. Because sea temperatures are much less variable and cannot be less than -1.9°C (see p. 4), coastal temperatures are usually milder than inland temperatures. The lowest recorded temperatures have therefore been collected from the centre of land masses at high latitude. In the northern hemisphere the lowest officially recorded temperature (-68°C) was reported from Oymyakon in easter Siberia, at least 400 km from the nearest coast and about 350 km south of the arctic circle. Paradoxically, it is appreciably warmer at the North Pole (situated on floating sea ice) than in northern Canada, Greenland or Siberia. The Antarctic situation is different, given the polar land masses and overlying ice-sheet, thousands of metres thick. The lowest 'ground level' temperature (-89.2°C) was recorded at the Vostok Base (USSR), nearly 1400 miles from the coast of Antarctica (and at an altitude of over 3000 m). It is the huge area of permanent thick ice that skews the temperatures of the southern hemisphere to lower temperatures than occur in the north.

The areas which have extremely low temperatures in winter also tend to have

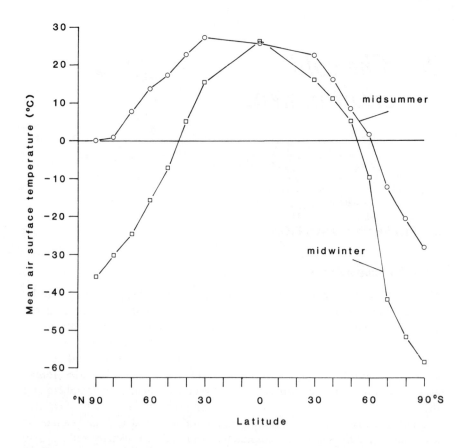

Figure 2.1 Mean midsummer and midwinter air surface temperatures at different latitudes. Note that the southern hemisphere is considerably colder, with less seasonal change in temperature than the northern hemisphere.

the greatest amplitude of temperature change between summer and winter. The 'cold pole' of eastern Siberia has been best studied in this respect. At Verkhoyansk, just north of the arctic circle, seasonal temperature changes between $-70°C$ and $+36.7°C$ have been unofficially reported — nearly 110 deg. C of temperature change. This temperature range also illustrates the fact that the freezing point of water is quite a high temperature for large sections of the terrestrial environment.

Why are high latitudes so cold, especially in winter? Basically there are three reasons. Firstly, radiant energy from the sun has to travel through a greater

(a) Attenuation of solar energy at high latitude

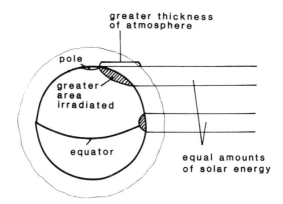

(b) Reflection of radiation by snow/ice

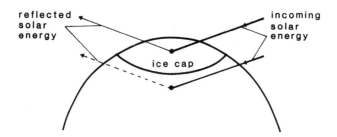

Figure 2.2 Reasons for colder conditions at high latitude. (a) shows that incoming solar energy (i) passes through a greater thickness of absorptive atmosphere and (ii) is spread over a greater area of the globe surface at higher latitude. (b) demonstrates that more energy is reflected back into space from white ice or snow covered areas. Thus ice caps tend to be self-perpetuating.

thickness of atmosphere to reach the ground (so is absorbed to a greater extent), and because of its oblique approach, the energy is spread over a wider area than is the case at low latitudes (Figure 2.2a). Secondly, because snow and ice cover land and sea at high latitudes, the reflectivity or albedo is high and incoming energy from a low sun is reflected tangentially away into space (Figure 2.2b). Thirdly, the tilt of the Earth's axis is such that, in winter, no solar energy reaches the polar regions for continuous periods of weeks or months (NB the arctic and antarctic circles are the lowest latitudes [66°32′ N and S] at which the sun does not rise on midwinter's day, or set on mid-summer's day).

Air temperatures are affected by altitude as well as by latitude. Temperatures fall by a daily average of about 1 deg. C for every 150 m of altitude increase. This altitudinal effect, which is independent of latitude, explains why there are mountains in the tropics which feature freezing temperatures on their upper slopes (e.g. Mount Kenya in East Africa, Cotopaxi in the Ecuadorian Andes). Latitude and altitude interact to make the upper parts of tropical mountains very demanding environments; profoundly cold at night, yet quite hot during the day. On the other hand, there is little seasonal change in tropical mountain conditions, whereas polar or temperate mountains show less diurnal change in temperature, but exhibit pronounced seasonal differences. This is particularly noticeable in the changing snow line of temperate mountain ranges such as the Alps and Tyrol. Mountains in general show much greater differences between sun and shade temperatures than do environments near sea level; this has implications for basking ectotherms such as lizards (see Chapter 3).

2.1.2 Sea temperatures

Sea temperatures have a much narrower range than air temperatures. Largely this is due to the great heat capacity of water (Chapter 1), combined with the enormous volume of the sea (about 500 million km^3). The heat capacity and volume interact to produce effective thermal buffering. Another reason for the restricted temperature range is that sea water begins to freeze at about $-1.9°C$, setting a lower limit to sea temperatures. Much heat of crystallization has to be extracted from water to create ice, so ice also functions as a buffer to temperature change. Surface oceanic water temperatures range between 30–35°C in the tropics in summer and $-1.9°C$ in the Southern Ocean and Arctic Ocean in winter. Tropical and polar surface water temperatures change relatively little seasonally (tropics approx. 5 deg. C, Arctic Ocean approx. 6 deg. C, Southern Ocean approx. 2 deg. C). The situation is different in temperate zones. Around the United Kingdom, sea temperatures may reach 19°C in September, yet fall to 2°C in January/February. Temperatures are more extreme in shallow seas; surface temperatures exceed 40°C in the Red Sea and Arabian Gulf; the Baltic freezes in winter despite a relatively low latitude.

Sea temperatures fall with depth. Virtually all deep water below 2000 m, whatever the latitude, has a temperature of 2–4°C (Figure 2.3). 80% of the area of the world ocean has a depth of more than 1000 m and the bottom water is formed of water cooled near the polar ice sheets until it sinks because of rising density (sea water has its maximum density at $-1.9°C$). The cold, dense water then flows towards lower latitudes beneath warmer, less dense water flowing towards the poles. The deep sea is made up predominantly of old, cold water which has not been touched by the sun's rays for centuries. From this picture, and from our knowledge of marine surface biomass levels (relatively low in the

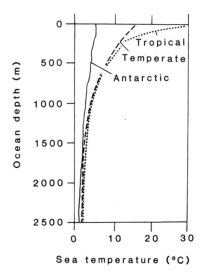

Figure 2.3 Changes in temperature with depth in the oceans. Note that temperatures are very similar at depths below 1500 m, whatever the latitude. Deep water is formed at the poles and has a temperature of between 2 and 4°C throughout the world ocean. Redrawn from Raymont (1967).

tropics, higher in temperate zones), it may be seen that most marine organisms spend most of their lives at temperatures below 5°C.

2.2 PALAEOCLIMATOLOGY

Palaeoclimatology relies on a battery of techniques, including traditional geological investigation, measurement of $^{16}O : ^{18}O$ ratios in the shells of fossil foraminiferans, study of palaeomagnetism and fauna/flora analysis. Broadly speaking, information about climate during the Quaternary (roughly the past two million years) is good and rapidly improving, but most data have been obtained for middle latitudes (currently cool temperate in nature). More attention is now being paid to the Tertiary, but inevitably detail is more difficult to obtain the further back in time investigation is pursued. For periods before the Tertiary it is really only possible to determine whether or not there is evidence of glacial strata. From Table 2.1 it may be seen that episodes of glaciation have occurred from the Precambrian onwards; Steiner and Grillmair (1973) recognize seven epochs of glaciation which are recognizable on several continents and which were separated by prolonged warm intervals. The causes of these epochs are essentially unknown, but numerous galactic and cosmological theories have been advanced (see Pearson, 1978 for review).

Table 2.1 Episodes of glaciation recognized by Steiner and Grillmair (1973). Note that each episode may have involved numerous ice ages. The ranges of age quoted are derived from observations in several continents

Glacial epoch	*Absolute age range (million years before present)*
Gowganda glaciation	2200–2460
Infracambrian II	900–1000
Infracambrian I	715–825
Eocambrian	560–680
Siluro-Ordovician	410–470
Permo-Carboniferous	235–340
Late Cenozoic	0–14

To return to consideration of more recent periods, where knowledge of climate and its causes is more satisfactory; during the Tertiary, average global temperatures were apparently stable and high, peaking in the Eocene (40–50 million years before present) at about 22°C, but steadily falling thereafter to a value of around 12.5°C at the beginning of the Quaternary (Nilsson, 1983). The Quaternary has been characterized by pronounced fluctuations in global temperature (Figure 2.4). These changes are now generally believed to be driven by the Lagrange/Croll/Milankovitch astronomical cycles which reflect:

1 changes in the shape of the earth's orbit (96 000 year periodicity);
2 changes in the earth's axis tilt from 21.5° to 24.5° and back (42 000 year periodicity)
3 equinoctial precession ('wobble') (21 000 year periodicity).

These cycles control the amount of solar energy reaching the earth's surface and, during the Quaternary, have interacted to produce cold glacials ('ice ages') and warm interglacials. During glacials the amount of ice present on

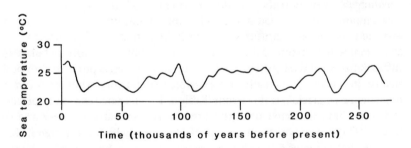

Figure 2.4 Fluctuations in Caribbean sea temperatures over past 250 000 years. Redrawn from Emiliani (1972).

land near the poles rises, sea levels fall and so do atmospheric carbon dioxide levels (for reasons which are not yet clear). A reverse sequence of events takes place during interglacials. The 'greenhouse effect' (produced by the release of CO_2, methane and chlorofluorocarbons into the atmosphere as a result of Man's activities), currently attracting much popular and political interest, is likely to produce a 'super interglacial' and delay the onset of the next ice age. The longest periodicity of the Milankovitch cycles appears to be most important and intervals between ice ages have been about 100 000 years. During glacials, global average temperatures have dropped below 0°C; currently we are close to the end of an interglacial with a global average surface air temperature of about 9°C. Global averages must be interpreted with care; equatorial temperatures have varied much less than the global average, while temperatures in temperate/boreal areas have fluctuated much more. However, Emiliani (1972) reported that Caribbean surface sea temperatures have been as much as 1 deg. C warmer in the past 500 000 years than the present 27°C, as well as being 5 deg. C colder. This implies air temperatures several degrees different from the present, even in subtropical regions. Within the present interglacial, temperatures have apparently been at least 2 deg. C higher in the Mediterranean area (approx. 7000 years ago).

The terms glacial and interglacial are relative; the Antarctic continent has been separated from all other continents for about 25 million years and appears to have been glaciated to a greater or lesser extent throughout that period, so in one sense we have been in an 'ice age' throughout that period; the Arctic appears to have been heavily glaciated only during the past two million years. Stonehouse (1989) remarks that the world is currently in an anomalously cold state and that temperate rather than cold poles have been the rule during most of the geological history. Although the astrophysical reasons for the Quaternary interglacial/glacial alternations in climate are becoming increasingly well understood, palaeoclimatologists can as yet provide no explanation for the prolonged warm periods, often lasting hundreds of millions of years, which have separated episodes of glaciation (Table 2.1).

Much information is also available about past sea levels, interpretation being based upon geological studies of fossil coral reefs in areas thought to have avoided geologically recent glaciation and tectonic activity (e.g. New Guinea), combined with analysis of $^{16}O : {}^{18}O$ ratios in the shells of fossil foraminiferans (which allow estimates of the relative volumes of water in the sea and in ice sheets). Sea levels have varied greatly during the past 160 000 years, from as much as 120 metres below present levels, during a glacial period (Shackleton, 1987), to 6 metres above current sea level. It should be recognized that sea levels are not simply determined by the amount of water locked up in ice sheets on land. The density of sea water is affected by temperature, so if the seas cool, they also shrink. It also should be remembered that formation or melting of sea ice has no affect whatsoever on sea level (melting ice cubes to not cause a gin and tonic to overflow!). Sea level changes are likely to have profound effects on animal life as they can destroy or create habitats, separate populations or bring them together. The current concern about the greenhouse effect has led to much more intense studies of the interaction between global temperature, weather, the sea and terrestrial ice sheets. At the time of writing it is becoming clear that the interaction is more complicated than previously realized. Sea level does not simply reflect a balance between the quantities of sea water and ice. Because the density of sea water changes with temperature, any warming will tend to cause the volume of the seas to expand, thus raising sea levels. However, warming of the air over the oceans tends to allow more moisture to be carried by the atmosphere, this moisture being precipitated as snow on the polar icecaps (thereby reducing sea levels!). Again, at the time of writing, the ice sheet covering the Arctic Ocean appears to be becoming thinner (but this in itself will have no effect on sea level), while the Greenland icecap is becoming thicker. Palaeoclimatology is a difficult enough field; predictive climatology seems fraught with chaotic uncertainty!

2.3 CLIMATIC ZONES

In popular parlance arctic and antarctic refer to those regions at higher latitudes than the arctic and antarctic circles. However, although these circles have precise geographical and physical meanings, they are of little use in bio-geography. To a terrestrial biologist, the Arctic is often defined as the area north of the natural treeline, since this will obviously be of some ecological significance. In Europe, as one travels north, deciduous woods are replaced by conifers and then by birch scrub. Finally trees (arbitrarily defined as woody plants of man size) disappear altogether ('treeline') to be replaced by tundra (marsh, grasses, sedges, moss and lichens). The treeline is also affected by altitude, and has often been changed by Man's activities too (conifers are often seen in northern Norway or in Iceland at latitudes which they would not normally survive — people plant them in sheltered sites and protect them from frost in the winter). Using the treeline definition, all of the Eurasian arctic is well within the arctic circle, and the terrestrial arctic is a relatively small area (<500 km wide in Siberia) on the northern coast of that land mass. In contrast the bulk of northern Canada is treeless, as are the whole of Iceland and Green-land, despite the fact that much of these areas are south of the arctic circle. All of the arctic is characterized by permafrost — permanently frozen subsoil at some depth below the ground surface.

An alternative definition of the arctic can be applied to both land and sea; it is that region north of the summer 10°C isotherm (i.e. the mean air tempera-ture of the warmest summer month). This results in a rather smaller terrestrial 'arctic' since the 10°C isotherm is normally about 100–200 km north of the treeline. Roughly speaking the 10°C summer isotherm corresponds to a 5°C summer sea temperature. Such sea temperatures are characteristic of waters influenced by warm currents such as the North Atlantic Drift (Gulf Stream) and there is often a sharp ecological distinction in water masses between polar waters (approx. 2°C) and much warmer waters near to the atmospheric 10°C isotherm (Figure 2.5).

The subarctic zone is even more loosely defined. Basically it is the area of coniferous forest (with some component of deciduous trees such as birch and

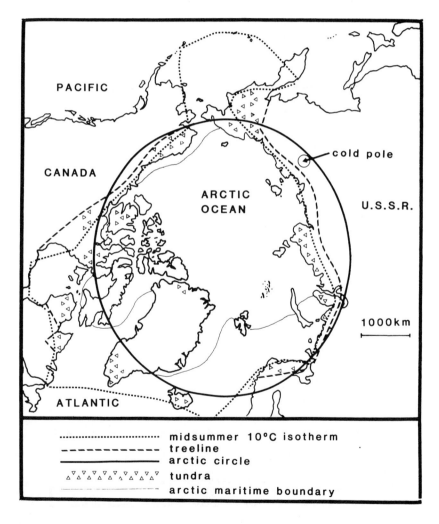

Figure 2.5 Arctic Ocean and surrounding land masses.

alder) which lies between the arctic and the temperate deciduous forests. In northern Europe these coniferous forests project well north of the Arctic Circle; in Canada they tend to be south of that circle. Permafrost is common and the subarctic zone includes the 'cold pole' of eastern Siberia, and the profoundly cold (in winter) regions of central Canada.

Definitions of zones in the southern hemisphere tend to be different. The continent of Antarctica is so large that portions of it protrude beyond the Antarctic circle (Figure 2.6). If the 10°C isotherm was used to determine the limits of the antarctic, then this would include the southern tip of South America, the Falkland Islands and South Georgia. Treelines are of no help either; Antarctica had deciduous woods in the remote past, but no trees have survived subsequent glaciation, which has substantially exceeded present

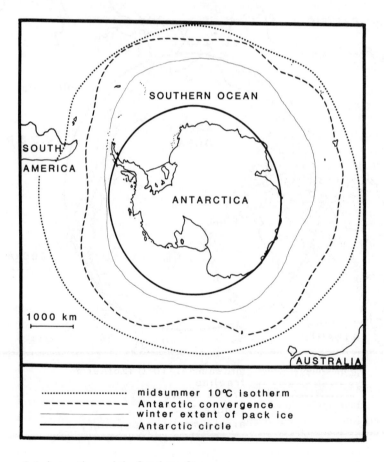

Figure 2.6 Antarctica and the Southern Ocean.

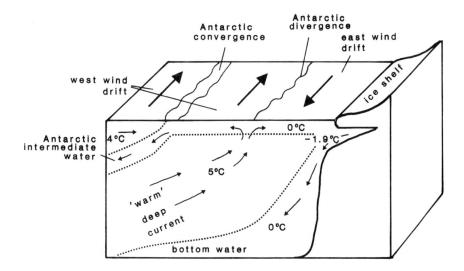

Figure 2.7 Patterns of circulation in the Southern Ocean.

levels even as little as 5000 years ago. Two maritime boundaries have tended to be used; the Antarctic Convergence of the Southern Ocean (Figure 2.7), found between 47° and 62°S (depending on season, weather and sector of the Southern Ocean), and the winter limit of pack ice (which tends to determine the severity of winter in the small islands around the antarctic land masses — those caught in the ice are much colder than those that remain ice-free and hence subject to the moderating influence of sea temperatures). Most biologists consider that the Antarctic Convergence constitutes the boundary of the Antarctic, which will therefore include South Georgia, the South Shetlands and the Southern Orkneys. The subantarctic zone includes the Falkland Islands, a number of small islands (including the Isles Kerguelen) known as the southern temperate islands (Stonehouse, 1982) and the tips of South America and South Island, New Zealand. None of this categorization is par-ticularly tidy; South Georgia, though regarded as antarctic is outside the pack ice, has a (relatively!) equable climate and a much more diverse vegetation than do the islands of the South Orkneys which have ice caps and are surrounded by ice during most winters.

Temperate zones are no easier to define than the polar regions. On land the temperate zones feature deciduous woodland — or at least they did before Man cleared forests to yield arable land. Leaf loss in winter is associated with pronounced seasonal light and temperature changes, so this definition has some ecological relevance. In popular parlance 'temperate' has overtones of mildness, but cold temperate areas can be extremely demanding in thermal terms with temperatures well below 0°C in winter. Large areas of northern Europe, Asia and America feature regular bouts of temperatures below −10°C, and temperatures are often much lower in particularly severe winters, sometimes reaching −30 or −40°C. Even in the UK, which has a reputation for a moderate climate, temperatures as low as −27°C were recorded in Shropshire in the mid 1980s.

Much of the area of tropical and subtropical regions of the world features sustained high temperatures on land; they will be little mentioned in this book. Exceptions occur in the case of low latitude deserts. Ecologists define deserts simply by rainfall (or snowfall at high latitude) and lack of trees. Treeless, sparsely populated areas with an annual precipitation below 255 mm are deserts whether they be in North Africa, Australia, Greenland or Antarctica. However, deserts of all types feature extremes of cold, even in subtropical and tropical areas. This is because the lack of vegetation and cloud cover allows much radiation of heat into the atmosphere at night. North African and Australian deserts can therefore feature freezing conditions (typically down to about −2°C) as well as temperatures above 50°C during the day. Naturally most attention has been paid to the problems of high temperatures as far as desert animals are concerned, but they have to cope with cold too.

2.4 HIGH LATITUDE MICROHABITATS

Except in the abyssal zones of the deep sea, habitats are made up of a mosaic of microhabitats, each with a different thermal regime. Many of these (e.g. burrows, polynyas) are exploited by animals as refuges to avoid exposure to extremes of temperature. Others are particularly stressful (e.g. mountain tops and permafrost layers) and avoided by most species. Examples are legion (see Stonehouse, 1989), but three will be highlighted here.

2.4.1 Polynyas

Polynyas are openings formed in sea ice (Figure 2.8). Temporary polynyas can form almost anywhere in pack ice, but there are a number of coastal sites in the Arctic and Antarctic where more long-lasting polynyas form on a regular basis because downslope (katabatic) winds drive newly-formed ice away from the coast. Such polynyas allow numerous birds and mammals to winter at high latitudes, or allow them to start breeding early in the spring before the pack ice has dispersed. Polynyas can be highly productive, because the thin, translucent ice which forms in them (before being blown away into the pack) acts as a substratum for rapid diatom growth.

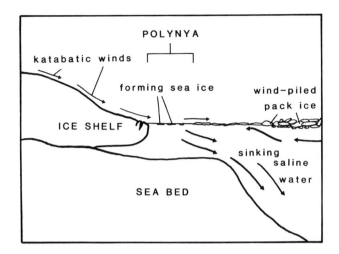

Figure 2.8 Characteristics of sustained polynas. Polynyas are areas of open water in pack ice. Most are ephemeral, but coastal polynyas form regularly in areas where katabatic winds flow down the ice cap and continually drive newly formed sea ice away into the pack. The formation of sea ice from sea water results in the formation of cold water of heightened salinity; this sinks, causing circulation within the polynya; conditions which favour high productivity in the spring.

2.4.2 Permafrost

The permafrost layer consists of permanently frozen ground that does not thaw in the summer. Permafrost starts at a depth of 1–1.5 m in arctic and subarctic regions, and its thickness varies with latitude. High in the arctic and antarctic permafrost is geographically continuous and may be up to 1000 m thick, the lower limit being determined by geothermal influences. Consequently it may penetrate for miles into the sea bed in shallow areas of the arctic basin. At lower latitudes, permafrost is discontinous (Figure 2.9) or patchy. Permafrost is important to animal life for many reasons. Firstly, it restricts drainage, so the soils overlying permafrost are often water-logged. Secondly, permafrost prevents animals from exploiting burrows as relatively warm refuges. The presence of permafrost is probably an important factor in preventing much of the subarctic from being colonized by hibernating amphibia, reptiles and small mammals.

2.4.3 Subnivean microhabitat

At high latitudes, thick snow is a normal winter feature, present for months on end and reaching depths (excluding drifts) of 1–2 m. Typically, the lowest soil

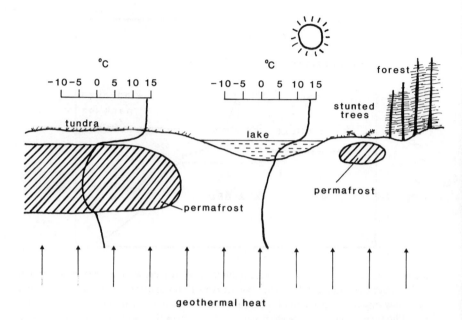

Figure 2.9 Semi-diagrammatic description of permafrost (permanently frozen soil layers). Note that permafrost layers cause soil temperatures above the permafrost to decline with depth; they also cause tree growth to be stunted. Any trees that do establish themselves above permafrost are prone to being blown over as their root systems cannot penetrate deeply enough to provide adequate support. Permafrost is not found beneath lakes, since the densest lake water overlying the lake bed cannot be colder than 4°C (the temperature of maximum density in fresh water). Permafrost cannot easily be established beneath evergreen coniferous forests which maintain relatively milder microclimates than the open tundra.

temperatures (well below 0°C) are recorded in the autumn, before snow falls. As soon as the snow is 20–50 cm thick, the poor conducting properties of snow reduce the amount of heat extracted from the soil by cold air, and geothermal heat flow from below warms the soil to about 0°C, so that the soil is unfrozen. Freeze–thaw cycles, plus local topography result in a space beneath the snow. If the snow is relatively soft, this space can be augmented by burrows dug by animals, mainly small mammals (Figure 2.10). The whole sub-snow environment is known as the subnivean microhabitat. Because Inuit and other peoples have many different words to describe different snow states, a rather complex terminology has been developed, mainly by North American scientists, to describe the fine features of the subnivean environment (e.g. Pruitt, 1984). However, the important features of this microhabitat are that the winter temperature range is much narrower (roughly 0 to −10°C if the snow tunnels are

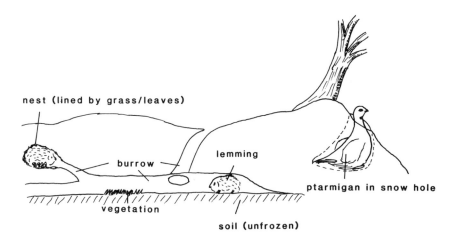

Figure 2.10 Subnivean habitat. Thick snow has poor thermal conductivity and acts as an insulating layer between cold air and the soil. This allows geothermal heat to warm the soil surface to around 0°C and provides a gap between soil and snow, soon augmented by the burrowing activities of small mammals (e.g. lemmings, shrews). Some vegetation remains green and edible; herbivores may also subsist on dried grass/leaves stored in summer. Larger animals (foxes, ptarmigan) can use snow holes as insulation.

included) than the air temperatures above the snow; that moving from snow tunnels to the soil itself gives access to the relatively warm temperature of 0°C; that light penetrates snow sufficiently to allow many plants within the subnivean zone to stay green and succulent (Salisbury, 1984), and that a variety of arthropod invertebrates (either active or inactive) live beneath the snow and provide food for soricids (shrews) (Aitchison, 1984). The subnivean habitat is mainly exploited by invertebrates and small mammals (e.g. lemmings, shrews and voles), but some larger animals (ptarmingan, foxes, hares) use the relatively warm environment beneath the snow as a refuge in harsh weather.

The subnivean environment is being recognized increasingly as the key to high latitude and high altitude survival for a range of small mammals. The equable nature of the microclimate is such that a number of mammals reproduce under the snow cover. Ecological considerations are complex, because of the food supply situation, but, from a purely thermal point of view, the snow cover means that the winter months are not the coldest for the subnivean communities.

PART TWO

Behaviour, Anatomy and Physiology

3 Behavioural responses to low temperature

Most observational and experimental study of adaptations to deal with low temperature has been directed towards anatomy, physiology and biochemistry. However, both endothermic and ectothermic animals often exhibit a range of behavioural responses which are effective in maintaining body temperature, or which allow animals to maintain exposure to cold conditions. Much convergent evolution of behaviour patterns has taken place and there are also no neat distinctions between the responses of ectotherms and endotherms. Accordingly, responses will be discussed under the broad headings of movement, gregariousness and sheltering.

3.1 MOVEMENT

3.1.1 Migration

Migration, the movement of animals from one area to another, is a feature of the life of a wide range of animals, from insects to elephants and from crabs to blue whales. The reasons for migration are rarely simple, but in many cases there are direct or indirect benefits in terms of movement to favourable thermal environments. Birds have attracted most study, not least because of the army of amateur (but expert) ornithologists who provide so much data about bird movements. During summer there is a tendency for many bird populations to move to higher latitudes where they exploit food resources to rear young during periods of retreating ice/snow, high productivity and relatively warm temperatures. Examples are legion, but waders, terns and duck species are conspicuous aquatic species, while martins and swallows (*Hirundo rustica*) are amongst the best-known terrestrial species (feeding on the summer abundances of insects). At the end of the summer, the birds retreat to lower latitudes. Arctic terns (*Sterna paradisaea*) even fly from one pole to the other. They nest north of the arctic circle in the northern summer (perhaps in Greenland or northern Norway), then fly 20 000 km to the edge of the pack ice of Antarctica, where they forage during the southern summer. Most of their life is spent in constant light, with virtually permanent access to summer productivity. During their long migratory flights (almost entirely over the sea), they rarely alight, but take small fish at the surface en route, never closing the wings.

Migration allows birds to exploit areas which would be lethal in winter, not so much because of low temperatures, but because foraging areas are covered by ice or snow, or are devoid of food plants and animals. This is particularly true of the smaller penguins of the Southern Ocean. Species much as the Adele and Chinstrap penguins (*Pygoscelis adeliae* and *Pygoscelis antarctica*) breed on the Antarctic Peninsula and nearby islands during the spring and summer. During the breeding season the adults and chicks feed upon krill (*Euphausia superba*) and upon small fish. At the end of the summer the penguins leave land and spend the winter at sea, north of the edge of the pack ice. By doing so they (a) avoid exposure to low air temperatures; (b) maintain access to food. If they stayed on the Peninsula they would have to cope with the problem of sea ice formation (totally covering the sea and therefore preventing swimming) and temperatures of $-30°C$ or lower. Occasional stragglers soon die, but of starvation rather than directly because of cold. For these small species, staying put throughout the year is not an option (in contrast to large species such as the Emperor Penguin *Aptenodytes forsteri* which overwinters and breeds on ice).

A few small high latitude birds are probably obligatory migrators because of cold alone. The snow bunting (*Plectrophenax nivalis*) is a small finch found north of the arctic circle in summer. At environmental temperatures below $-40°C$ the core body temperature of the bird starts to fall. Snow buntings cannot survive arctic winters, yet even in this case it is probable that day length or food supply trigger migration, and not cold *per se*. Migrations of terrestrial mammals, such as those of reindeer and caribou appear to be driven by similar forces; the animals can survive high latitude temperatures, but cannot reach their diet of lichen if the snow cover is too thick. In Canada the herds of caribou spend the winter dispersed in forests. They migrate northwards about 400 km onto the tundra in March and April. Calves are born in May and June when the snow has melted and plenty of mosses and lichen are available. Wolves, dependent on young and sick caribou for food, migrate with them. Both predator and prey lay down fat in the summer to fuel the return migration and eke out the poor winter food supply.

The migrations of marine mammals are usually performed for similar reasons. Some seal species (e.g. the Ross (*Ommatophoca rossi*) and Crabeater (*Lobodon carcinophagus*) seals) are adapted to maintain ice respiration holes which permit more or less permanent high latitude life, but other pinnipeds (e.g. Walrus

(*Odobenus rosmarus*), Elephant Seal (*Mirounga leonina*)), must remain at the edge of the sea ice or they risk being trapped below the ice (unable to breathe), or trapped above it (unable to feed). Mortality due to these causes is not unknown; elephant seals hauled out at Signy Island, South Orkneys occasionally fail to enter the sea before sea ice forms — they invariably die of starvation as their poor terrestrial locomotory abilities do not allow them to crawl quickly enough to reach the edge of the advancing ice sheet. Remaining near the edge of the sea ice can mean annual migrations of hundreds of kilometres as the ice expands and shrinks. However, some populations of walrus in the Canadian Arctic do not have to migrate — they survive winter as far as 80°N in large 'recurrent' polynyas (Chapter 2) that are found predictably each year because of hydrographic features which prevent ice formation (some polynyas are known to have existed each year since the 17th century).

Whale migrations have attracted much study. Toothed whales (e.g. the killer, *Orcinus orca*) often stay at high latitude in the winter (though they usually remain outside the pack ice). In contrast, the baleen whales make long migrations to subtropical waters in the winter. Particular study has been devoted to the Blue Whale *Balaenoptera musculus*. The whales are in Antarctic waters from December, when the krill density is high, until March, when it is declining (a total period of about 120 days). The whales then leave the Antarctic and move to subtropical waters (particularly off Africa) where they spend the rest of the year (Lockyer, 1981). They therefore rely on stored energy for about 245 days per year (Mackintosh, 1966). Calves (usually single) are born and grow in warm waters, before moving to the Antarctic with the mother, so much of foetal and neonatal development is fuelled from energy stores too. Kshatriya and Blake (1988) have recently considered migration energetics in blue whales, and have demonstrated mathematically that, although extremely large blue whales (approx. 200 000 kg — about the maximum size reported) might remain within their thermoneutral zone in Antarctic waters at −1.9°C, most members of the population (especially young animals of around 10 000 kg, but even mature whales of 100 000 kg) have to expend energy above basal levels to maintain core temperatures. Figure 3.1 shows the effect of body size on lower critical temperature (LCT) in the species; it also demonstrates that the smallest blue whales need to be at latitudes of less than 20°S if they are not to expend energy in maintaining body temperature. Kshatriya and Blake calculated that there were energetic advantages in migrating (at optimum cruising speeds) to lower latitudes when food was unavailable. Their model predicts that smaller animals should migrate further north than large whales, and also indicates that young animals ought not to venture as far south as mature animals when feeding on krill. There is some observational support for these hypotheses from Matthews (1978), who reported that immature blue whales had a more northerly distribution than older animals throughout the year. He also noted that the pigmy blue whale (a smaller subspecies) did not forage as far south as its larger

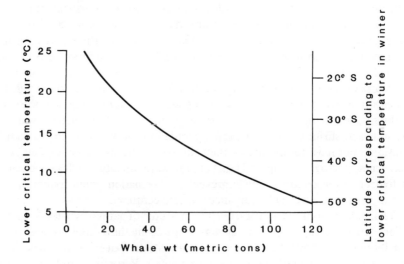

Figure 3.1 Lower critical temperatures for blue whales (*Balaenopterus musculus*) of various sizes. These are the temperatures at which whales will have to increase energy expenditure to maintain core body temperatures. Latitudes corresponding to the critical temperatures are also indicated. Redrawn from Kshatriya and Blake (1988).

relative. However, much smaller baleen whales (especially the minke, *Balaenoptera acutorostrata*) do forage at the edges of the pack ice, indeed the minke has been recorded in winter polynyas hundreds of kilometres within the pack ice (Stonehouse, 1989). It should also be remembered that whales are social animals, that may gain feeding and reproductive benefits from living in small groups ('pods'), and so may to some extent trade off the advantages of thermal neutrality against the advantages of social existence. Nevertheless, the work of Kshatriya and Blake is important since it demonstrates that migration is necessary for all blue whales on thermal grounds, not simply because new born animals need a warm environment.

Consideration of temperature-driven migrations in ectothermic animals seems to have been limited mainly to aquatic species. Terrestrial ectotherms either have severely restricted locomotory ranges (e.g. amphibia, reptiles), or they overwinter in a dormant state (many insects, some amphibia). Migrations have been considered in marine species though. A number of mobile intertidal species migrate out of the intertidal zone as winter starts and return in the spring. Examples noted in Europe include the shore crab (*Carcinus maenas*) and the intertidal teleost, *Blennius pholis*. Not all shore crabs leave, but only small specimens remain; these are probably small enough to burrow into the substratum and to inhabit crevices so that they are not exposed to the full rigours of cold air when the tide is out. Some blennies also remain on shore during the winter months, but they are limited to relatively deep rock pools which retain heat from the sea for long periods at low tide. It is doubtful whether the crabs or blennies migrate in response to falling environmental temperatures; they probably react to shortening day length, which allows them to 'anticipate' cold weather, a form of predictive behavioural thermoregulation (see below). In the southern ocean there are few intertidal animals, but one of them, the limpet, *Nacella concinna*, which is the only large invertebrate to be found between the tidemarks south of latitude 60°S, moves into subtidal waters in the autumn, staying there until the pack ice retreats in the spring.

Many species of pelagic fish are very mobile, capable of swimming hundreds or even thousands of kilometres. In both hemispheres there is a tendency for fish to move into deeper water during winter. At high latitudes this may protect them from exposure to the risk of freezing (by keeping them out of surface waters laden with ice crystals), but the reason for this migration may again be concerned with food; most pelagic fish rely on a diet of zooplankton, and zooplanktonic organisms tend to move out of surface waters as phytoplankton production dies away in the autumn.

In summary, therefore, migrations often transfer animals from one thermal environment to another, and their movements may take place along thermal gradients. However, there appears to be relatively little evidence that temperature itself triggers migration, or that the immediate prize of migration is a warmer environment.

3.1.2 Movement along small-scale thermal gradients

Migrations usually take place over relatively large distances, beyond those susceptible to laboratory experiment. As we have seen, animals move along thermal gradients when they migrate, but they do not need to sense temperature to do so. This type of behavioural thermoregulation is known as predictive thermoregulation (Neill, 1979) and relies on the predictable nature of the animals habitat. This can apply to small scale movements too, and can have a learned component, at least in vertebrates. Neill suggested that freshwater fish might remember the location of warm springs and move towards them in cold weather. They would not need to be capable of detecting or swimming along a thermal gradient. Let us take the example of a fish in river which has a warm spring 1 km upstream of the fish's current location. On a cold day in winter the ambient temperature (t_a) falls below the preferred temperature of the fish (t_p). To move to water of a higher temperature, the fish needs to swim upstream until $t_a = t_p$. To do this by predictive thermoregulation it needs (a) to recognize that $t_a < t_p$; (b) 'know' that a warm spring is upstream of its present position; (c) recognize the upstream direction and swim along it and (d) recognize when $t_a = t_p$ (near to the spring).

Predictive thermoregulation on a local basis is probably more widely used by terrestrial animals, particularly endotherms, because discontinuous thermal regimes are more prevalent on land. If a lizard or locust moves into a sunny area when its body temperature is low, and retreats into shade when its core temperature is high, then it achieves the same result whether it is using visual cues, or following thermal gradients. In most circumstances movement between light and shade is likely to be quicker if visual cues are employed. Movement of mammals (including man) into sheltered situations, caves and suntraps often involves an instinctive or learned predictive element. Predictive thermoregulation, whether instinctive or learned, has considerable advantages, particularly in energetic terms, because it will generally ensure the most direct and rapid movement between unsatisfactory and preferred thermal environments. A sheep which moves directly from the centre of a field to the shelter of a wall when the north wind blows is behaving more efficiently than it would if it had to try to detect and track faint and highly variable thermal gradients in the atmosphere swirling across the field. However, there can be serious problems arising from reliance on predictive thermoregulation, which is rather inflexible — an animal relying wholly upon predictive thermoregulation has no defence against a breakdown of the predicted relationship between temperature and space. Such breakdowns are perhaps easier to accomplish in the laboratory than to observe in the field, but in the example of the freshwater fish given above it is evident that, if the warm spring fails, the fish may die as it cannot seek out unknown sources of heat. A laboratory demonstration was provided in a study upon the intertidal gammarid amphipod *Gammarus duebeni* by Davenport (1979b). On subarctic shores in northern Norway the animals live in rockpools which freeze over during winter low tides. The amphipod normally responds to falling temperatures by swimming downwards into colder water of increasing salinity, thus avoiding being frozen by the thickening surface ice. In the laboratory Davenport managed to cool a vessel of water in such a way that the water at the bottom froze first; the gammarids still swam downwards in response to falling temperatures and were promptly frozen solid!

Some animals are certainly capable of detecting and moving along thermal gradients. A variety of terrestrial arthropods (e.g. woodlice) will move along a thermal gradient to select preferred temperatures, while pythons and vipers in search of endothermic prey (usually small rodents) can track them in complete darkness by use of thermal receptors (though they are not thought to use such behaviour for thermoregulatory purposes). A particular group of animals which appear to move along temperature gradients are the ectoparasites (lice, fleas, mites, bugs) of endothermic animals. All wild birds and mammals are infested with such animals, which exploit the warm air within the fur or plumage of their hosts to maintain relatively high body temperatures. In cold weather these ectoparasites tend to accumulate in the warmer regions of the skin (axillary and inguinal regions in mammals; beneath the wings in birds). Such behaviour may also direct them to regions of the body where their food (blood) is more accessible, and may help them to transfer from one host to another (Davenport, 1985b). Murray (1976) found that lice (*Lepidothirus macrorhini*) of the southern elephant seal (*Mirounga leonina*) accumulated in the tail flipper area when the animal was 'hauled out' and basking in the sun with the flipper blood vessels dilated. This behaviour kept the lice in a favourable thermal enivronment (27–34°C). The lice of elephant seals maintained in captivity under constant temperatures were evenly distributed over the body. Newborn mammals have also attracted some study; they are often blind, yet need to be able to locate their mother or siblings for suckling or huddling. Rabbit kittens (12–72 hours old) have been shown to move along thermal gradients to reach areas of favourable temperature (Satinoff *et al.*, 1976), congregating at an optimum temperature of 36.4°C. This ability is probably widespread in young birds and mammals which have thermal receptors in the skin.

There is a third form of behavioural thermoregulation by movement which involves neither predictive relationships, nor the ability to detect and follow thermal gradients. This behaviour, which depends upon responses to recent thermal experiences, is known as reactive thermoregulation and has been studied in teleost fish, though the principle could be applied to any mobile ectothermic animals. In 1954 Sullivan noted a great difference between the activities of fish at given temperatures depending on their recent thermal history. She found that thermally acclimated fish (i.e. fish held at constant temperature for long periods) were most active at the species' preferred or optimum temperature. This finding is as expected from a range of studies on ectothermic animals. However, Sullivan found that fish which had been exposed to changeable temperatures were least active at the temperature preferendum! This result was confirmed for salmon (*Salmo salar*) by Ivlev (1960) who noted that in fish exposed to heating or cooling rates of 0.1 deg. C. min^{-1}, the swimming speed was proportional to the difference between ambient temperature and preferred temperature (Figure 3.2). These observations led to a simple model which suggested that randomly swimming fish in a heterothermal environment (i.e. one which contains water at a variety of temperatures) would finally aggregate in water of the preferred temperature because swimming activity is at a minimum in such water. Over subsequent years this research field has become highly mathematical, and analysis has shown that the simple model is faulty. Unless fish become motionless at the preferred temperature, they will (given a limitless thermal environment) eventually stray by chance into areas of lethal temperature. This theoretical problem is averted if a turning component is introduced, with fish tending to turn more often the greater the difference between ambient and preferred temperatures. Thus, as they travel towards areas of lethal temperature, they will tend to turn more often, but will turn less frequently when approaching water of preferred temperature. There are many refinements of this basic model; the more mathematically inclined should consult Neill (1979).

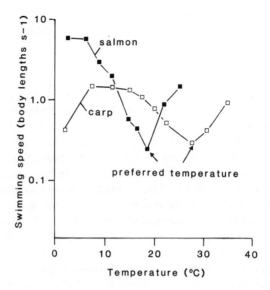

Figure 3.2 Swimming speed of carp and salmon in relation to changing temperature (redrawn from Ivlev, 1960). The fish were exposed to heating or cooling rates of 0.1 deg. C min^{-1} during the course of the trials.

3.2 BASKING

Most terrestrial and freshwater habitats are heterothermal in space and time. On land the surface of rocks and vegetation provide exposure and shelter as far as solar radiation is concerned, while the surface waters of calm lakes and slow-moving rivers are warmer than the depths. Weather and season affect these temperature differences, with winter tending to abolish them (cloudy weather, turbulent mixing of water), while summer sun accentuates the differences. As far as land is concerned these differences tend to disappear fairly rapidly at night (because of thermal radiation into the atmosphere), but temperature differences are often maintained in freshwater habitats because of the great thermal capacity of water. Many ectotherms take advantage of temperature differences in their habitat to maintain body temperatures within a favoured range. Endotherms also bask, but to control the energetic cost of thermo-regulation rather than body temperature itself.

3.2.1 Ectotherms

Examples of ectothermic basking are well known; many species of lizards and snakes alternate basking in the sun and hiding in the shade to maintain high and steady body temperatures; this was first noted by Cowles and Bogert (1947) who found that a lizard could have a cloacal temperature of 38°C despite an air temperature of only 13°C. Such reptiles are known as shuttling heliotherms. Thus it is possible to find the adder (*Vipera berus*) and viviparous lizard (*Lacerta vivipara*) throughout subarctic northern Norway, hundreds of kilometres north of the arctic circle. Because the sun is above the horizon throughout the summer they can bask to maintain temperature at high levels for many hours each day (in good weather). They can therefore sustain the activity necessary for efficient foraging and reproductive behaviour. They must also accumulate lipid reserves for the winter period when behavioural thermoregulation becomes impossible, so is spent torpid in burrows below levels affected by frost. Lizards and snakes are also distributed as far south as Tierra del Fuego; only Canada, Greenland and Siberia (plus a few islands such as Iceland) are devoid of either. The common European lizard, *Lacerta vivipara*, has been much studied by Roger Avery and his students at Bristol University. Despite living in a cold temperate climate with a yearly average temperature of little more than 9°C common lizards have a preferred body temperature (PBT) of 30°C. They have an active temperature range (in which hunting, feeding and fighting are all possible) of about 28–32°C. If cooled to 25°C they cannot feed efficiently. Basking is very effective; in summer they warm from 15 to 25°C in about 5 minutes. When they reach body temperatures in the activity range, they stop basking and start foraging (Avery, 1979). When the body temperature falls to the lower level of the activity range, basking is restarted. For several hours each day their body temperature is maintained within the activity range, but, because they are small animals there is a rapid fall to ambient air temperatures as soon as basking ceases in the early evening (Figure 3.3). For *Lacerta vivipara* basking is only feasible between about March and September (in the UK — the season will be shorter in more northerly populations) and then only when sun is available. At this point it is worth noting that both of the common northern European reptiles (*Lacerta vivipara* and *Vipera berus*) are viviparous. Although both species overlap with oviparous species in the southern part of the range, it seems probable that viviparity combined with shuttling heliothermy allows the mother lizard or adder to act as a sort of mobile incubator, speeding development of the young.

Lizards can even survive at high altitude in the Andes by virtue of basking. Pearson and Bradford (1976) studied basking in the lizard *Liolaemus multiformis* at an altitude of 4000 m in Peru. At night the air temperature falls well below 0°C, but the lizard survives by retreating into its burrow. Basking starts early in the morning (Figure 3.4) at abut 7.00 am when the air temperature may be as low as −5°C. The lizard lies in the rays of the newly risen sun on a mat of vegetation which insulates it from the surrounding ice and snow. Within two hours the body temperature has risen to 35°C (i.e. similar to that of endothermic birds and mammals), and can be kept at that level for most of the

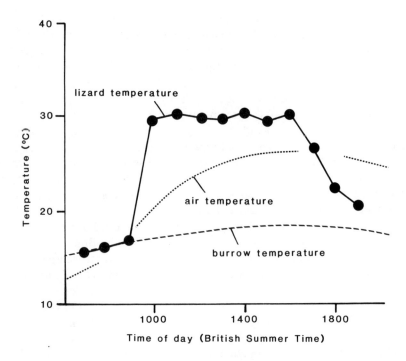

Figure 3.3 Mean body temperature of three specimens of the viviparous lizard *Lacerta vivipara*. The animals were in an outdoor enclosure and could retreat to a burrow. When they left the burrow in the morning, basking in the sun soon allowed the body temperature to rise to 30°C, far above both air and burrow temperatures. Basking permitted maintenance of high and stable temperatures for nearly 6 hours. Redrawn from Avery (1979).

day unless the weather is cloudy. The high body temperature permits the lizard to move rapidly enough to catch its invertebrate prey, and also promotes quick digestion. Speed of movement is particularly important to predators and behavioural thermoregulation plays a part in sustaining that speed; Greenwald (1974) showed that the velocity of the strike of a gopher snake (*Pituophis melanoleucus*) was directly related to the animal's body temperature, and that an equally direct connection existed between strike velocity and success in catching prey. Basking allows snakes and lizards to maintain 'rates of living' comparable with endotherms for part of their lives, without requiring the sustained high metabolic rate and food input of those forms.

Such behavioural thermoregulation probably has a very long history judging by the results of studies upon the tuataras (*Sphenodon punctatum*) of South Island, New Zeadland. Tuataras superficially resemble lizards, but are the sole surviving representatives of the order Rhyncocephalia, an order which, like

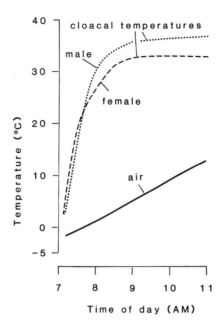

Figure 3.4 Body temperature changes during the day in a high altitude lizard *Liolaemus multiformis* (redrawn from Pearson (1954)). Two lizards were studied; there appeared to be no significant difference in thermoregulatory ability between the sexes. Despite the high altitude and low air/ground temperatures these lizards maintain a higher day time body temperature than the viviparous lizards shown in Figure 3.3.

the snakes and lizards, is descended from Triassic lepidosaurs. Rhyncocephalians date back some 220 million years ago, and the tuatara is very similar to fossil rhyncocephalians of 140–170 million years ago. The remaining tuataras live on a few cold temperate islands off the shore of New Zealand. They are mainly nocturnal and burrow dwelling and normally have low body temperatures (6–16°C). However, tuataras do bask at the mouth of their burrows on warm days (Bogert, 1953). Saint Girons (1980) studied *Sphenodon* in rather more detail. He noted that some tuataras made no effort to bask, whereas others sought out sunlit clearings and shelters and basked actively, thereby achieving temperatures as high as 25–26°C. Further study revealed that the basking tuataras had consumed large meals during the previous nocturnal foraging session (whereas non-basking animals had not fed). Presumably the heightened body temperature promotes rapid digestion. Since the basking animals have a body temperature at least 10 deg. C. higher than the non-baskers, it is probable (assuming a Q_{10} of about 2) that the digestion rate is doubled. Saint Girons also suggested that male tuataras (and probably snakes and lizards too) may bask to promote spermatogenesis during the breeding season. It is evident from these studies that ecological 'costs' and 'benefits' have much influence on basking behaviour; it 'pays' a recently fed tuatara to raise its body temperature to digest the meal, converting the ingested material into assimilated protein, lipid and carbohydrate. Once digesion is complete it 'pays' the tuatara to resume low environmental temperatures to minimize loss of energy.

Basking is not limited to vertebrates; many insects bask too. Hocking and Sharplin (1965) found a most attractive example of basking at high altitudes; a number of mosquito species raise their body temperatures by spending much of their time in the parabolic blooms of arctic poppies. The poppies themselves orientate the flower towards the sun during the day, so the mosquitoes stay warm enough to fly in search of warm-blooded prey despite air temperatures which would make them torpid. Williams (1981) reported that the butterfly *Euphrydryas gillettii* of the Wyoming mountains uses shrubs (*Lonicera involucra*) for basking, but in this case temperature is controlled in the next generation! The female butterfly lays her eggs on the highest large leaves of the shrub, but preferentially on the underside of leaves which face southeast and consequently intercept the morning sun. Measurements show that these leaves are the warmest and that eggs laid upon them develop and hatch earlier than elsewhere. Williams also presented convincing evidence to show that the rapid onset of winter in Wyoming means that a faster egg development rate allows the subsequent larvae to feed for a longer period and therefore have increased fitness.

Basking is also seen in aquatic ectotherms. The alligator (*Alligator mississipiensis*) occupies warm temperate as well as subtropical areas. It will move into shallow warm water to raise its body temperature, but also makes quite sophisticated use of atmosphere and sun (Smith, 1975, 1979). Alligators first expose part of the skin of the back to air and let it dry ('pre-basking posture'). If the air temperature is high enough, the animal will crawl out of water to bask in the sun. This behaviour suggests that the alligator possesses thermal receptors in the skin. Reverse-basking (cold-seeking) is also seen in some aquatic ectotherms. Blennies (*Blennius pholis*) seek out deep cold water in rock pools during the summer (Davenport and Woolmington, 1981), while Brattstrom (1979) reported that larval and adult frogs shuttled between warm and cold water.

Basking is often quite a refined process, not a simple matter of sitting in the sun. One of the most remarkable and complex pieces of thermoregulatory behaviour is that demonstrated by the Galapagos Marine Iguana, *Amblyrhynchus cristatus*. This large lizard is well adapted structurally and physiologically to its intertidal and subtidal habitat, where it feeds upon seaweeds, often at depths of 10 m or more. The Galapagos islands are on the equator, so might be expected to be outside the scope of this book. However, the islands are bathed by surprisingly cold currents, so the iguanas are exposed to temperatures of 20–25°C when they dive. Dives last for several minutes and the iguanas gradually cool down, despite well developed peripheral vasoconstriction (presumably under the control of a central 'thermostat', perhaps in the hypothalamus). After a dive the Galapagos Marine Iguana basks in the sun to raise its body temperature as rapidly as possible to the PBT of 36°C. The foreshore on the Galapagos islands consists in the main of unshaded black lava flows, and the rock temperature may exceed 50°C on sunny days. Initially the cold iguana flattens its body against the rock and orientates the body at right angles to the sun's rays to expose as much of the body to the sun as possible. However, when the body temperature has risen to about 40°C (presumably pushing the metabolic rate up by a factor of 3–4), *Amblyrhychus* has to take action to avoid overheating (Bartholomew, 1966). The iguana orientates the body parallel to the axes of the sun's rays and raises the head and shoulders so that most of the body and tail are in shade, and the forepart of the body is off the substratum, exposed to cool sea breezes. This behavioural thermoregulation is crucial to survival; tethered iguanas soon overheat and die.

Shuttling heliothermy requires neurophysiological control of behaviour. The animal concerned needs to regulate its temperature by means analagous to the physiological control systems of endothermic birds and mammals (Chapter 4), but with shuttling largely replacing peripheral vasomotor control. A lizard which demonstrates behavioural thermoregulation needs (a) to recognize that its temperature is changing (up or down); (b) to be capable of identifying the direction of a warmer or colder climate; (c) to be able to move rapidly in the appropriate direction. In addition, the animal needs to have a reference temperature range within which non-basking behaviour occurs. This last point is important; a single set temperature (acting like a thermostat switch) would result (via negative feedback) in permanent shuttling back and forth between warm and cold environments! The animal also needs to 'decide' that the sun has gone in (or down) and that behavioural thermoregulation should be abandoned *pro tem*. Evidence from a variety of lizard species leads to the picture of control shown in Figure 3.5. Hardy (1976) demonstrated that there is a hypothalamic centre which is temperature sensitive in the brain of the blue-tongued skink (*Tiliqua scincoides*). The hypothalmus of the skink contained cold sensitive neurons (which fire more frequently when cooled), heat sensitive neurons (which fire more frequently when heated), together with neurons which are relatively temperature-insensitive. Berk and Heath (1975) obtained much data from the American desert iguana (*Dipsosaurus dorsalis*) using a shuttle box, cold at one end and warm at the other. They demonstrated that the iguana had two set point temperatures, one of which triggered cold-seeking, one of which triggered heat seeking (Figure 3.5). Berk and Heath envisaged motor neurons (M) supplied with inhibitory and excitatory synapses. The excitatory synapses would be activated from hot or cold sensitive neurons (H,C) in the hypothalamus. The inhibitory synapses would be activated from

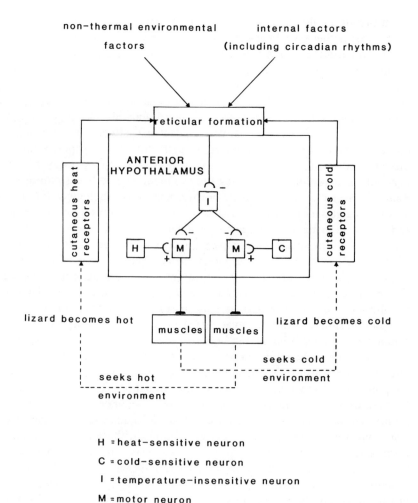

Figure 3.5 Simplified layout of the neural network of the thermorgulatory control mechanism of the desert iguana, *Dipsosaurus dorsalis*. Synapses are represented as stimulatory (+) or inhibitory (−). (Modified from Berk and Heath, 1975).

thermally-insensitive neurons in the hypothalamus, themselves controlled by cells in the reticular formation, which in turn are sensitive to hot and cold sensors in the skin. Berk and Heath's model (slightly modified here) proposes that a lizard basking in the sun will warm up. The heat-sensitive neurons will begin to increase firing rate, tending to stimulate the excitatory synapse on the surface of the motor neuron. At the same time, the hot sensors in the skin will increase their firing rate, leading (via mediation in the reticular formation) to inhibition of the temperature-insensitive neurons of the hypothalamus. This causes a reduction in firing rate along the axon leading to the inhibitory synapse on the surface of the motor neuron. As excitation increases and excitation declines, a threshold will eventually be reached when the motor neuron fires, initiating muscular action which helps to transfer the animal to a cooler part of the environment. The animal will then cool, reduced output from skin heat sensors and hot-sensitive hypothalamic neurons will cause the motor neurons which control cold-seeking behaviour to cease firing, and the animal will stop shuttling behaviour. Cooling will continue until rising output from cold-sensitive hypothalamic neurons and skin cold sensors shifts the balance of excitatory and inhibitory output to the motor neurons controlling heat-seeking behaviour sufficiently to result in firing of those motor neurons.

This part of the model allows the lizard concerned to have a body temperature range over which foraging and intraspecies activity take place. Further refinement is needed however. Shuttling needs to be 'switched off' either at night, or seasonally in the case of animals living at relatively high latitudes. This could be done by increased firing rate of the temperature-insensitive (inhibitory) neuron (mediated via the reticular formation), so that its inhibition could not be countered by the maximum firing rate of the hypothalamic cold receptor. Switching off could be done in response to input (excitatory or inhibitory, depending on neuronal layout) from the optic centres of the brain (which could integrate light intensity over a period of time), or the pineal 'eye' could be involved, particularly in the case of seasonal switching. The pineal or third eye appears to function as a radiation dosimeter (Avery, 1979) and has been implicated in many seasonal phenomena in reptiles, particularly those living at higher latitudes. A deficiency of the model lies in the lack of detail about translation of motor neuron output into movement in the appropriate direction; again environmental information delivered to the reticular formation from peripheral thermal and visual sensors should suffice.

Behavioural thermoregulation by shuttling heliothermy not only confers physiological and biochemical benefits (resulting from the maintenance of body temperatures closer than would be the case in the absence of regulation), it also results in physiological or ecological costs. For example; the shuttling behaviour demands the expenditure of energy in locomotion (a physiological cost). Such movements may also make the animal conspicuous to predators (an ecological cost to a population if the lizard is consumed), as may mere presence in some exposed basking sites — a lizard basking on a rock is more visible to a bird than a lizard foraging in the undergrowth. Also, time spent in thermoregulatory behaviour is largely unavailable for other activities such as foraging or reproducing. An extreme example of this situation is seen in the high altitude Andean lizard *Liolaemus multiformis*. Pearson and Bradford (1976) found that specimens living at 4500 metres spent 82% of an average 24 hour period deep in their burrows (thereby avoiding freezing). 12.3% of their time was spent in basking in the sun and 3.5% in picking up heat by flattening themselves against warm parts of the substratum ('thigmothermy'). This left only 2.1% of the day for social activity, travelling or feeding!

All of the 'costs' of behavioural thermoregulation, either to the individual or to the population (whether by shuttling heliothermy, migration or any of the other mechanisms described in this chapter) reduce the effective physiological/biochemical benefits:

$$\text{gross benefit} - \text{costs of thermoregulation} = \text{net benefit}$$

In cases where the costs of behavioural thermoregulation exceed gross benefits then the animal concerned will be better off energetically if it becomes passive with respect to temperature. Huey and Slatkin (1976) have considered the question of lizard thermoregulation in great detail, but were predominantly concerned with modelling mathematically the behaviour of tropical/subtropical species. They were able to demonstrate that some lizards (e.g. the small Puerto Rican iguanid *Anolis cristalellus*) inhabited both 'low cost' and 'high cost' habitats. Low cost habitats were open parkland, where basking sites were numerous and involved little movement to reach them. High cost habitats were forested, with few clearings. In parklands the little anoles maintained stable temperatures (by shuttling between sun and shade) for much of the day, only becoming thermally passive at night, but in forested areas their body temperatures were always close to ambient air temperatures.

Similar principles can be applied to ectothermic animals basking at higher latitudes, but in this case the costliness of habitats will vary with season and weather (more variable than the sunny tropics). Prolonged periods of cloudy weather in summer will be as effective as the short daylength of winter in inhibiting behavioural thermoregulation. Lizards and snakes will have to spend much of their life in a thermally passive state, often at temperatures so sub-optimal that feeding and digestion cannot take place. In such species summer heliothermy must generate enough heightened activity for the species to collect sufficient excess energy to permit lipid storage to tide the animals over subsequent cold seasons or prolonged bad weather.

3.2.2 Endotherms

Basking in endotherms has attracted little quantitative study. Many birds and mammals living in cool or cold climates spend part of their time basking in the sun. By doing so they are unlikely to raise core body temperature (unless they spend too long in direct sunlight), but basking will permit peripheral tissues to operate at temperatures closer to the core value than is the case under cold conditions. Such basking will obviously tend to keep the endotherm concerned within its thermoneutral zone, thereby reducing energy expenditure. Basking seems to be particularly important to aquatic endotherms (e.g. seals, penguins), which lose heat when they dive (even though efficient peripheral vasoconstriction permits long term maintenance of core temperatures. Seals and penguins may often be seen with their flippers extended at right angles to the sun's rays, maximizing heat uptake. Elephant seals raise their tail flippers off the ground and spread the webs wide on sunny days. Eventually basking may be so effective that the seals have to lose heat. Fur seals (both northern and southern) have been seen to sweat from the back flippers (which are also waved around to create a flow of air over them) when too hot. If sweating is impossible, basking spells may be ended by the bird or mammal retreating to shade or cold water.

One of the few quantitative investigations of solar energy input to an endotherm was carried out by Øritsland (1970). He calculated that a polar bear (*Thalarctos maritimus*) basking in full sunlight picked up energy roughly equivalent to the basal metabolic rate. This indicates the scale of energy saving possible — and also explains why Norwegians living around Tromsø (well inside the arctic circle) are able to sunbathe out-of-doors in minimal swimwear during early May (provided they surround themselves with windbreaks of snow!).

3.3 GREGARIOUSNESS

Huddling or clustering behaviours are important in permitting survival in cold conditions. The principal is simple; if a number of warm animals huddle together so that much of their surface area is in contact with other animals rather than the cold environment, then the effective surface area exposed to the environment is much decreased, heat loss is slowed and the animals will save energy or may even be able to maintain a body temperature which would not be sustainable if the animal was alone. All other things being equal, the value of huddling increases with the difference between body and ambient temperatures $(t_b - t_a)$, and the effectiveness rises with the number of animals forming the huddle (Madison, 1984). Although gregariousness has usually been implicated in behavioural thermoregulation of endotherms, it is also of value to some groups of ectotherms.

3.3.1 Endotherms

All small mammals sleep and rest in contact with other individuals when they can. Such huddling is crucial to the survival of young mammals (and nestling birds too) which have imperfect physiological thermoregulation. However, huddling is also important to adult animals. Gregarious behaviour is employed by numerous small birds to reduce energy demands at night or in cold weather. House Sparrows (*Passer domesticus*) and starlings (*Sturnus vulgaris*) roost together in huge numbers at night (the author has seen a single tree in Sussex where as many as 15 000 starlings were to be found together every evening). Particularly interesting are the responses of wrens (*Troglodytes troglodytes*). Wrens are amongst the smallest of cold temperate birds, reach no more than 8 cm in length, weigh perhaps 30–40 g and spend most of their life amongst vegetation or rocks so are rarely noticed. It is generally accepted that they are the most common of British birds, with around 10 million pairs breeding in most years. Wrens are ubiquitous given cover, living on islands around the UK coasts, yet are also found on mountain tops. They are vulnerable to prolonged cold weather (the population fell by 75% in the long, severe winter of 1962–63) because of food and water shortage. On the other hand, they are well adapted to deal with cold *per se* because of their well developed huddling behaviour. The degree of gregariousness is remarkable; Soper (1982) reported that 61 wrens had been observed to enter a nest box of 11.4 × 14 × 14.6 cm at night under cold conditions. This equates to a space per bird of only 38 ml. Taken with the weight range given above, this suggests that the nest box was effectively full of wrens (weighing about 2 kg in all), with little room for air circulation. The degree of energy saving in this circumstance would have been tremendous, particularly as the huddling was combined with shelter. However, there are costs as well as benefits; one of the birds died during the night, perhaps of trampling. Large aggregations of roosting birds also attract predators.

Pearson (1960) was one of the first workers to investigate benefits of huddling quantitatively. He studied the harvest mouse *Reithrodontomys megalotis* and found that if three mice were placed together at a temperature of 1°C, then their oxygen uptake was only 72% of that recorded when they were held separately at the same temperature. Harvest mice, like other small mammals, require food at frequent intervals for survival. A 28% reduction in metabolic rate achieved simply by clustering in such small numbers makes a great contribution to saving of foraging time (particularly as foraging in cold weather is itself costly in energetic terms). Part of the food collected during foraging expeditions has to pay the locomotory and handling costs of foraging, so the saving in terms of food collection will substantially exceed 28% in this case. Studies of the type carried out by Pearson are open to the criticism that single animals are psychologically stressed by their isolation, that their metabolic rates are likely to be elevated, and that the experimental design tends to overestimate the energetic value of huddling. This point was addressed (and dismissed) in a careful study by Contreras (1984) who investigated the metabolic rate of (a) single animals; (b) separated trios (in sight of their conspecifics) and (c) huddled trios. He worked on both white mice (*Mus musculus*) and Mongolian gerbils (*Meriones unguiculatus*) and found that isolated animals used no more oxygen than individuals of separated trios, yet individuals of huddled trios used substantially less oxygen than when separated

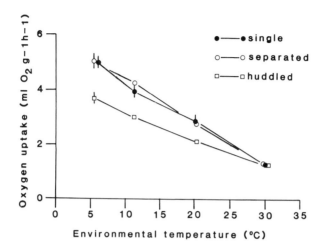

Figure 3.6 Rate of metabolism of Mongolian gerbils (*Meriones unguiculatus*) at different environmental temperatures. 'Single' represents the mean oxygen uptake of three gerbils measured separately, out of sight of one another. 'Separated' indicates uptake by three gerbils in sight of one another but unable to maintain body contact; 'huddled' represents uptake by the same three gerbils allowed to maintain contact with one another. The results demonstrate that the gerbils do not show elevated oxygen uptake when held singly, so the lowered metabolic seen in huddled animals is a real thermoregulatory response. Redrawn from Contreras (1984).

(Figure 3.6). As with the harvest mouse, there was an energy saving of about one third associated with huddling. Contreras also established that the absolute energy saving increased with falling ambient temperature (Figure 3.7), and that there was relatively little advantage associated with huddling in groups larger than three animals (Table 3.1). He believed that this was largely due to

Table 3.1 Reduced oxygen uptake in huddled white mice at a temperature of 12.5°C

Number of animals in huddle	$\dfrac{Oxygen\ consumption\ huddled}{Oxygen\ consumption\ separated}$ × 100
1	100
2	82.1 ± 2.4
3	69.1 ± 1.2
4	67.5 ± 4.9
5	67.3 ± 5.9
6	67.7 ± 2.0

After Contreras, 1984.

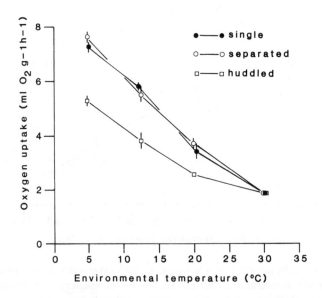

Figure 3.7 Rate of metabolism of white mice (*Mus musculus*) at different environmental temperatures. 'Single' represents the mean oxygen uptake of three mice measured separately, out of sight of one another. 'Separated' indicates uptake by three mice in sight of one another but unable to maintain body contact; 'huddled' represents uptake by the same three mice allowed to maintain contact with one another. These data are presented to show that the thermoregulatory benefits of huddling increase with decreasing temperature. Redrawn from Contreras (1984).

huddles being two dimensional, so that a cluster of small mammals has a different shape from a single animal.

Bats also cluster apparently inefficiently in thermoregulatory terms, since clusters have to be two dimensional as the animals hang from surfaces. However, Kunz (1982) suggested that clustering cut oxygen uptake by half at temperatures below the thermoneutral zone. More recently Kurta (1985) investigated the insulating properties of bat clusters by using dead specimens of the little brown bat (*Myotis lucifugus*). He found that clustering in groups of seven bats (the outer six surrounding a single animal) resulted in an improvement in insulation of 103% — a value compatible with Kunz's assertion. A combination of the shape, posture (huddling bats tend to wrap their wings around themselves) and position at the top of caves and attics is likely to contribute to the greater energy saving of clustering in bats by comparison with mice/gerbils.

Most studies of the value of winter huddling in small mammals have been concerned with active animals with a normal body temperature. However, Arnold (1988) has recently investigated social thermoregulation during hibernation

in alpine marmots (*Marmota marmota*). Alpine marmots are rodents that hibernate in large underground dens (roughly 60 cm × 60 cm × 60 cm). Usually
parents plus one or two year classes of offspring hibernate together in groups
of some 5–8 animals. Hibernation is discussed elsewhere (Chapter 5) in detail,
but here it is necessary to appreciate that marmots alternate long periods
(about 2 weeks) of torpor and low body temperature (typically about 8°C) with
short periods of arousal and normal body temperature (36°C). Arousal, during
which body temperature rose from torpid to normal values was quick (occupy-
ing about 7 hours); normal body temperature was usually maintained for about
24 hours. Repentry into full torpor took about 5 days. Arnold found that
the cycle of torpor, arousal, normality ('euthermy') and return to torpor was
closely synchronized within a group (Figure 3.8). The animals always huddled
together when they were euthermic, thereby gaining the thermal benefits
described above, but they also usually huddled when torpid. When the first of
the animals aroused at the end of a bout of torpor, some of its body heat was
transferred to other members of the huddle, thereby warming them passively,
and reducing the amount of heat that they have to produce during arousal.
When the first animal was completely aroused and euthermic it lay beside or on

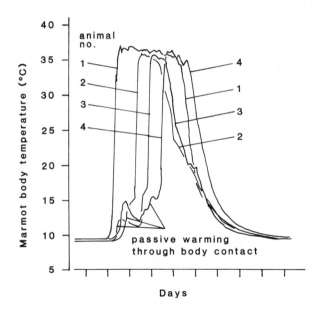

Figure 3.8 Illustration of synchronized arousal and warming in a huddling group of
hibernating alpine marmots (*Marmota marmota*). Note that animals 2, 3 and 4 gain some
of their warming from body contact with animals already aroused. Redrawn from
Arnold (1988).

the torpid animals, groomed them and also covered them with nest material — all actions contributing to energy saving during their arousal. Synchronization is also advantageous during re-entry into torpor as it would be difficult for an animal to reduce its body temperature effectively in the presence of warm nest mates.

Clustering during hibernation is particularly important to one group of small mammals, the bats. Bats are basically of tropical origin. Most are insectivores and food is extremely scarce during temperate zone winters. However, bats cannot burrow or construct nests. Bats often cluster in aggregations of several hundred animals (see Davis, 1970 for review). Although the clusters are two-dimensional (p. 70), they occur at the top of enclosed spaces (e.g. caves, attics) so retain heat efficiently. Clustered hibernating bats are able to maintain body temperatures of 5–10°C despite surrounding air temperatures of −10°C or below, though many species choose to hibernate in areas of cave systems which do not drop much below 5°C. Clustering in overwintering bats is not designed to maintain normal body temperatures; that would be counterproductive as the bats need to reduce their metabolic rate during the winter to survive. Instead, clustering allows bats to maintain a low (but safe) body temperature without risk of freezing. In considering the importance of gregariousness of chiropterans it must be remembered that bats cannot fatten up before winter to the extent feasible in hibernating ground dwellers — such weight increases are not compatible with flight! Clustering ekes out precious lipid reserves.

Huddling may also be of some thermal value to larger endothermic terrestrial animals living at high latitudes and exposed to low temperatures. Several social animals have been observed to huddle, notably wild horses, musk oxen (*Ovibos moschatus*) and penguins, but such huddles should not immediately be assumed to conserve body heat. The primary purpose of huddles may be to guard against predation (as in the tight circles of musk oxen, effective against wolves). In addition, unless the animals are sheltered from winds it is probable that air movements will strip away heated microclimates. However, two of the larger penguin species, the King and Emperor penguins (*Aptenodytes patagonica* and *Aptenodytes forsteri*) which bred on level areas rather than distinct nesting sites, are believed to gain thermal advantage from huddling (as originally determined by Stonehouse, 1953). The emperors in particular occur in breeding colonies which may contain thousands of birds (each up to 45 kg in weight and about 120 cm in height); they also breed in the winter when temperatures may fall to −50°C and wind speeds approach 180 kph. Emperors demonstrate several aspects of gregariousness. The male birds each carry a large egg (0.5 kg) on the upper surfaces of their feet, the egg being in contact with the naked surface of a fold of skin which acts as a pouch. Egg and pouch are largely covered by the parent's feathers. This intimate contact ensures that the egg is maintained at a high incubation temperature (close to 40°C) despite brutal air temperatures. Next, when conditions are fairly cold the emperors tend to form small groups of semicircular form, each bird with its back to the wind. This helps to protect the eggs (or chicks after hatching). Finally, under severe conditions the emperors form tight huddles ('tortues' in the French literature — named after the Roman 'testudo' (tortoise) military formation of tightly packed troops). The tortues have to be formed on a flat surface, usually sea ice (Jouventin, 1975) — when emperors are inland during the early stages of egg incubation they remain in smaller groups and exploit the sheltered sides of hilly terrain. Tortues can sometimes consist of thousands of tightly packed birds and probably save as much as 80% of heat loss (the tall, cylindrical shape and lack of ears, together with a short tail undoubtedly contribute to a greater energy saving than is possible in huddles of small mammals). It is worth emphasizing that adult emperor penguins have body weights which overlap with the body weights of adult humans (albeit small ones). Anyone who has been part of a crowd of football spectators on a cold day will appreciate the thermal advantages of tortues, and even the most tightly packed stadia do not approach the crowd densities of *Aptenodytes*! The minimum size for adult tortues in the worst weather appears to be about 300 birds, and all of the existing emperor colonies are larger than this. Emperor chicks, and the chicks of other penguins, particularly Adélies, also form tortues once they have entered the creche phase. These tortues may be extremely tightly packed and somewhat igloo-shaped as the peripheral animals lean in towards the middle.

As is so often true in the case of behavioural thermoregulation, there are costs associated with penguin huddling. Disruptive squabbling often stems from close proximity and pushing by peripheral birds. Sometimes penguins fall and are trampled. Large tortues contain a considerable biomass of bird flesh (perhaps over 100 tons), occasionally enough to overstress thin sea ice. If the birds do fall into the water they will not come to any harm — but eggs and chicks will be lost, representing a considerable ecological cost to the population and a waste of the months of energy expended by the parents.

Gregarious behaviour is also important to another group of high latitude animals, the pinniped seals and walruses. Pinnipeds are well adapted to life in cold water; they can maintain steady core temperatures indefinitely in water with a temperature between -1.9 and $+2°C$. Problems arise however, when they haul out onto ice or land and are faced by low air temperatures ($-10°C$ or lower). All are relatively hairless (the short fur of adult Weddell and Fur seals is of little insulative benefit) and will gradually lose heat when temperatures are low or winds strong. A few hours of exposure seem to be tolerable, but eventually pinnipeds retreat to water to 'warm up'. This response is inappropriate in some cases because of circumstances which prohibit a return to water. The southern elephant seal *Mirounga leonina* is a case in point. This seal is one of a few species which undergo a 'catastrophic moult'. Fur is lost in patches and the seals remain out of water virtually continuously for a month. Clumsy on land, they are likely to generate little heat by muscular thermogenesis and gradually use up fat — though it has to be said that little visible impression is made on the blubber of the average 4 ton adult male which may have a girth equal to his length of 6–7 metres! The moult occurs in spring or early summer, so temperatures on South Georgia and Signy Island (where they have been most studied) are usually fairly equable. However, most of the elephant seals hauled out on Signy Island are to be found in groups (known as 'pods' by BAS personnel) of tightly huddled cows and young males, each made up of perhaps 10–30 individuals. Usually the pods are close to the water, but are sometimes formed in shallow depressions some distances from the sea. Pods commonly have an adult bull or two associated with them and undoubtedly have a warm, wet and smelly microclimate around them which melts the surrounding snow. Since a large pod may contain 20–30 tons of seal flesh with a limited effective surface area, this is perhaps to be expected, but the palpable warmth as one nears a large pod is still remarkable.

Pierotti and Pierotti (1983) pointed out that young seal pups have a temporary furry coat which does allow them to survive solo exposure to cold and wind when they are being suckled and incapable of swimming. The pups' parents face a dilemma in deteriorating weather. Male seals move off the ice when air temperatures drop below −20°C. Females continue to feed pups, but leave the ice as soon as suckling has finished. Walruses (*Odobenus rosmarus*) face a different situation in the breeding season; their pups are hairless and vulnerable to cold. Whenever the weather is poor, the mother walrus remains out of water in intimate contact with her offspring.

A special case of the thermal importance of gregariousness is one of the most familiar — the incubation of eggs by birds. Bird egg development is very sensitive to temperature and, at least until late in the developmental period, the bird embryo is incapable of producing significant metabolic heat. The average incubation temperature is about 34°C, and the heat needed to sustain a temperature difference to as much as 30–40 deg. C must be provided by the parent(s). In some species both parents brood the eggs, but in those birds in which the male is brightly coloured and the hen rather dowdy, it is the latter which does all of the sitting. The brooding birds moult their down feathers from a well vascularized 'brood patch' of skin on the underside, and it is through this patch that heat is supplied to the clutch of eggs. At high latitudes, particularly in the southern hemisphere where nests are often rudimentary or constructed of stones, there is a tendency towards small egg clutches, and this may be partly due to the metabolic demands of keeping the eggs warm. Sitting birds respond to egg cooling by increased insulation (feather erection) and increased shivering thermogenesis (Tøien *et al.*, 1986; Aulie and Tøien, 1988), but it is probable that they rely on their own thermoregulatory responses (a cooling egg will extract more heat from the parent) rather than sensing the egg temperature directly.

3.3.2 Ectotherms

Thermal benefits of gregariousness in ectotherms have relatively rarely been considered except in the case of what may be described as 'social endothermy' in some insects which live together in large numbers. Resting insects are ecto-thermic (flying insects often have high body temperatures by virtue of the biochemical reactions of flight muscles, so may be described as exercise endo-therms (Schmaranzer and Stabentheiner, 1988)), so most insects at higher latitudes must pass the winter in some sort of dormant state (often as eggs or pupae). Some social insects living in nests, holes in trees or rock cavities are more active in winter, the best studied example being the honey bee *Apis mellifera*. Honey bees penetrate subarctic areas, yet become chilled and essentially moribund if the body temperature falls below 7.5°C (Butler, 1974). During the winter, individual honey bees regularly leave the hive to defaecate ('cleansing flight'), even when ambient temperatures are near freezing. They are able to do this because of muscular thermogenesis, the flight muscles liberating large quantities of heat. However, the bee is in a precarious state during such flights and if it lands on the ground, or lingers at the hive entrance it will quickly die. So how does the species overwinter at high latitude? It has been known for many years that honey bees form clusters of near-perfect spherical form during winter; these clusters may contain thousands of bees in large hives; the bees continue to feed throughout the winter. Clustering occurs when the temperature within the hive falls to around 13°C. Measurements within clusters have revealed stable temperatures of at least 20°C in early winter when the bees are without brood (i.e. eggs, larvae). In the late winter (February in the UK) breeding starts and the inner cluster temperatures, recorded from the zone where the new brood is held, rise to 32–35°C! Clearly the clustering bees have a great deal of independence of environmental temperatures outside the hive, but how do they achieve this state, so desirable from the point of view of survival and rapid development of the new brood of workers? The source of this independence was once thought to result from muscular thermogenesis (by wing fanning and abdomen wagging). This idea is superficially attractive, and wing fanning is certainly used to raise the hive temperature when the bees are not clustered (Michener, 1974). However, this idea is now regarded with some suspicion (at least for large clusters) since it would require considerable energy expenditure by the bees in muscular or locomotory movements at a time of the year when food supplies cannot be replaced.

It is more likely that bee clusters become warm and maintain high tempera-tures because of the low effective surface area of the cluster. Like all small ectotherms, the individual bee has a high surface area-to-volume ratio, and the relatively low resting level of metabolic heat production is rapidly dissipated to the atmosphere across the body surface. However, in a cluster of several thousand tightly packed bees, most of the insects are not directly exposed to the atmosphere, and the cluster becomes a single, large, near-spherical social endotherm (Davenport, 1985b) because heat no longer being dissipated to the atmosphere becomes available for raising the cluster temperature. Of course the cluster also needs the physical protection of the hive structure to minimize convective heat escape and to avoid heat loss by windchill.

Davenport (1985b) considered the effectiveness of clustering in the following idealized manner. Rates of transfer, whether of heat or of respiratory gases, between an organism and its environment tend to be proportional to the surface area of the organism rather than its volume or weight (see Zeuthen, 1970 for review). Thus, for example:

$$r = a\,(w)^b$$

where

r = respiratory rate

w = weight

and a and b are constants for a given species

if respiratory rate is proportional to surface area $r = a\,(w)^{0.67}$ from geometric considerations.

if heat output is proportional to respirary rate then

$$h = k(w)^{0.67}$$

where h = heat output

and k is a constant for a given species.

Consider a population of 10 000 'bees':

let each 'bee' have a heat output

$$h = k(w)^{0.67}$$

∴ 10 000 separate individuals will have a heat output of 10 000 h

However, if 10 000 bees are packed tightly together into a sphere then the cluster heat output

$$h_c = k(10\,000w)^{0.67}$$

where h_c = heat output across surface of cluster

$$\frac{h_c}{h} = \frac{k(10\,000w)^{0.67}}{k(w)^{0.67}}$$

∴ $h_c = 479\,h$

So, a cluster of 'bees' will behave thermally as if it consisted of only 479 individuals; a heat loss reduction of 95.2%. This saved heat will therefore be available to raise the cluster temperature. Obviously this is a highly idealized situation! The b is usually closer to 0.8 than 0.67 for multicellular organisms (Zeuthen, 1970); this would reduce the saving to about 84%. Also, the bees

Table 3.2   Theoretical advantages of clustering of bees

	% of metabolic heat available to raise cluster temperature	
Number of 'bees' in cluster	(1) b = 0.67	(2) b = 0.80
1	0.0	0.0
5	41.2	27.6
10	53.0	36.9
50	72.4	54.2
100	78.1	60.2
500	87.1	71.1
1000	89.8	74.9
5000	94.0	81.8
10 000	95.2	84.1
50 000	97.2	88.5

After Davenport, 1985.

need air spaces to permit gaseous exchange and this will further increase the cluster surface area and heat loss. On the other hand, as may be seen from Table 3.2, substantial benefits are gained by quite small clusters. Even 100 'bees' will theoretically have 60% of their heat output available for raising cluster temperature (b = 0.8). Interestingly Lecomte (cited in Butler, 1974) reported that clusters did not form in experimental cages if fewer than 50 bees were present, but formed within a few hours if 100 or more animals were available.

Southwick (1982, 1985) has investigated real bee clusters in great detail. He found that bees in relatively small clusters exhibited density-dependent mass-specific oxygen uptake rates. This indicated that bees clustered in small numbers have heightened metabolic rates (probably produced by enhanced thoracic muscle activity). In larger clusters the average metabolic rate per individual declined, being virtually stable at all cluster sizes above 600 g (approx. 4800 bees). This supports the suggestion above that bees in large clusters do not rely on sustained muscular activity for heat production. Southwick also showed that the metabolic rate and thermal conductance of bee clusters in winter is remarkably similar to those of birds and mammals of similar weight, reinforcing the concept of bee clusters as social or colonial endotherms. Additionally, he demonstrated that there is a gradient in bee temperature between the core of the cluster and the periphery (analogous to the temperature gradients measured in mammalian bodies). Southwick employed numerous thermocouples in his investigations of bee clusters. The output from these indicated that the bees within a cluster are in almost constant motion. He was also able to confirm the existence of a positional cycle, in which cooler bees migrate from the periphery to the warm core of the cluster before they became comatose; the minimum temperature of peripheral bees was about 9°C, just above the cold-coma temperature. Finally, Southwick introduced the idea that the thoracic body hair of bees may contribute to heat conservation. Bee hairs are short (<1 mm) and would have negligible insulative value for individual bees. However, Southwick (1985) envisaged the hairs interlocking with those of neighbouring bees to form insulation as effective as the fur and feathers of mammals and birds. The author feels that Southwick may well be pushing analogies a little far in this respect. Birds and mammals maintain a virtually motionless body of air within their pelage; a bee cluster will need to ensure enough air circulation to support respiration.

Bees control cluster temperature by altering the degree of packing. In relatively warmer conditions, the cluster is loosely formed, but becomes more compact as the temperature drops, reaching a minimum size when the external temperature falls to 0°C. Several factors appear to attract honey bees together to form clusters. Experiments have shown that they can move along thermal gradients towards warmer areas, so more bees would be attracted to an already functioning cluster. Bees are highly social animals, so are attracted to one another by a variety of olfactory (and possibly auditory) signals; such attraction, perhaps triggered by falling environmental temperatures, would help to initiate clusters.

Although the honey bee *Apis mellifera* exhibits perhaps the most remarkable winter temperature control of any ectotherm, it should be noted that other members of the Hymenoptera are able to control the physical conditions within their hives or nests during cool weather. Bumble bees (*Bombus* spp.) are able to keep nest temperatures up by fanning and heightened activity as long as there are large numbers of workers and plenty of honey to support them (Hasselrot, 1960). Hornets are also known to cluster at night to maintain high brood temperature (and optimum growth) (Ishay *et al.*, 1967). Indeed, the efficiency of thermoregulatory behaviour in wasps and hornets is nearly as impressive as that of the honey bee (see Spradbery, 1973 for review), though less commented upon in general texts, probably because their social organization collapses at the beginning of winter and only the queens survive (by low temperature hibernation). There is also good evidence that the wood ant, *Formica rufa* maintains heightened nest temperatures by virtue of metabolic heat (Brian, 1977).

Gregarious behaviour is found in other ectothermic groups, particularly terrestrial reptiles. Specimens of large constricting snakes (the Indian python *Python molurus*, the anaconda *Eunectes murinus* and the boa *Boa constrictor*) have all been found to cluster together at night in the cooler portions of their ranges (Myers and Eells, 1968). This behaviour apparently slows loss of the heat picked up by basking in the sun during the day — boas certainly lose heat more slowly when clustering than when held individually (in laboratory experiments).

Related to clustering is the special case of coiling in snakes. Several large snakes have been shown to gain thermal advantage by coiling. Thus, as shown by Cogger and Holmes (1960), the carpet snake *Morelia spilotes* picks up heat during the day by basking and then maintains an elevated body temperature during the subsequent cooler night by coiling tightly (which reduces the effective surface area by 50–70%). Such heat conservation may hasten digestion and aid in the incubation of eggs, but should not be confused with the brooding behaviour of the Indian python *Python molurus* which appears to rely upon muscular thermogenesis and an enhanced metabolic rate (as well as coiling) to raise the temperature of female and clutch (Hutchinson *et al.*, 1966).

Most other reports of clustering behaviour in ectotherms appear to be associated with huddling during hibernation. Several workers have reported winter aggregations of snakes and lizards. A good example is that of the European adder, *Vipera berus*. The most northerly distributed of all snakes (it may be found on the Kola Peninsula, north of the arctic circle), it often occurs in numbers up to 800 in single dens (Viitanen, 1967). In Finland, there is a tendency for increasing size of aggregation as one reaches higher latitudes, maximum numbers within dens being recorded at 63°N. At even higher latitudes, the numbers fall, but adders become dramatically scarcer anyway beyond this latitude. Whether such aggregations have direct thermal benefits is not known, and it is possible to make a good case for winter aggregations being necessary to facilitate mating immediately after emergence, so that the subsequent short spring and summer may be devoted as much as possible to basking and foraging, rather than for locating mates. Experimental investigations are obviously desirable to clarify the function of such aggregations.

Finally, there may be thermal benefits for gregarious intertidal animals. Many intertidal animals possess mechanisms which ensure that they settle amongst others of their species. Usually it is thought that such behaviour enhances breeding success either in sessile species which need to be close enough to copulate (e.g. acorn barnacles), or in the case of many invertebrates which broadcast their gametes into the environment (e.g. tubiculous polychaetes, oysters, cockles, mussels). Intertidal gregarious settlement helps to place organisms in good areas for food capture, and may help in the avoidance of predation. However, gregariousness can also have thermal benefits at subzero temperatures where freezing is a possibility. Intertidal animals are usually exposed to freezing conditions for a few hours at most. The minimum temperature reached, and the proportion of body water frozen is dependent to some extent on the size of the animal concerned; a large animal will cool more slowly than a small animal, especially when ice formation starts. It would be predicted, therefore, that a number of small animals in intimate contact with one another would tend to lose heat more slowly than would the same number of animals of similar size held individually. Davenport and Shaw (unpublished data) have recently tested this hypothesis upon the mussel *Mytilus edulis*.

Mussels held in groups survived better at freezing temperatures than did animals held individually. Mussel spat are usually found either closely packed together (as sometimes happens when they colonize virgin rock), or attached amongst larger mussels. This behaviour may have a variety of functions, but it does improve freezing resistance too.

3.4 SHELTER

3.4.1 Burrowing

A highly effective method of behavioural thermoregulation involves burrowing or the utilization of existing burrows. Thermal stability increases amazingly quickly with depth into the substratum (Figure 3.9). Turnage (1939) demonstrated that, although the temperatures at the soil surface of deserts varied by more than 80 deg. C during the year, the variation at a depth of 100 cm was only 12 deg. C. Deserts may be hot during the day, but they can be bitterly cold at night, so some fairly unlikely groups of animals may spend much of their life underground, thereby avoiding thermal extremes. Many

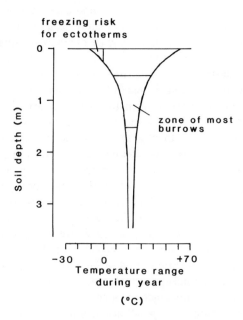

Figure 3.9 Changes in temperature with soil depth. Based on the desert recordings of Turnage (1939). Note that temperatures of burrows of depths greater than 1 m are virtually independent of season and time of day.

desert birds are burrow-dwellers, while in American deserts there are tortoises which spend much of their life underground. The gopher tortoise, *Gopherus agassizii* is a medium-sized chelonian (adults are more than 30 cm in length) and uses two sorts of burrows. 'Summer holes', dug in flat regions of the desert are no more than 1 m deep and serve to ameliorate diurnal temperature extremes. 'Winter dens' are permanent burrow systems in foothills at the edge of the deserts, often tens of metres in length. There is some evidence that winter dens are thousands of years old, being too extensive for single animals to dig; they may therefore be inhabited by generation after generation. Winter dens protect gopher tortoises against freezing during hibernation, provided that they are at least 5 m inside the burrow system (Pritchard, 1979). Cloudsley-Thompson (1970) reported somewhat similar behaviour from the Sudanese desert tortoise (*Testudo sulcata*). The same author also demonstrated that burrowing is crucial to the survival of young *Crocodylus niloticus* which burrow in cool weather (quite common in some areas of the Nile crocodile's distribution e.g. Zimbabwe).

Burrowing is also important to the lives of endotherms, particularly at high and temperate latitudes. Burrows can be into earth, snow or even be formed as runways of arched and woven vegetation. Life in burrows can make small mammals substantially independent of the external environment. Hayward (1965) monitored the burrow temperatures of six geographical races of deer mice (*Peromyscus*) living in habitats ranging from the Nevada desert to alpine meadows in Wyoming (height approx. 3500 m; mean annual air temperature −2°C). He did so by ingeniously inducing the deer mice to carry thermistor probes down into their burrows, but it must be stressed that the temperatures recorded were unaffected by the animals themselves which moved to other parts of the burrow system after releasing the probes. The depth of burrow was variable but there was always 1–1.5 metres of cable between the exterior and the probe. In all deer mouse habitats there was remarkable stability of burrow temperature despite great external fluctuations. Thus, in the Nevada desert, external temperatures fluctuated diurnally between 16 and 44°C, yet the burrow microhabitat had a temperature close to 26°C at all times, allowing the deer mice to be in their thermoneutral zone. Burrow temperatures were even more stable in winter, especially when snow cover augmented soil insulation. Deer mouse burrows in alpine Wyoming never showed temperatures below 0°C despite external temperatures as low as −15°C. Obviously huddling plus nest building can bring burrow temperatures well above 0°C. From his measurements Hayward was able to show that the temperatures encountered by deer mice changed very little with external environment, and that this explained the lack of structural or physiological differences between the geographical races.

Burrowing is also employed by numerous birds, particularly sea birds living in cold temperate and high latitudes. In the Southern Ocean, populations of the smaller petrels (e.g. Wilson's storm petrels (*Oceanites oceanica*)) and snow petrels (*Pagadroma nivea*) breed in rock cavities on scree slopes and in burrows beneath rocks. A number of petrels living in the Southern Orkneys (e.g. prions, *Pachyptila* spp.) nest deep in the peat of the moss billows which cover much of sheltered areas of the islands. In the northern hemisphere auks (e.g. puffins) and shearwaters behave in similar fashion, gaining thermal buffering for themselves, their eggs and young as well as protecting nestlings against predatory gulls and skuas.

Particularly interesting is the situation of the two genera of lemmings (made up of about a dozen *Lemmus* spp., *Dicrostonys* spp. living in the circumpolar tundra). Lemmings are small rodents of hamster size and spend the whole year in the arctic tundra. Like most arctic endotherms they do not hibernate. The winter is spent in snow burrows below 1–2 metres of snow ('subniveal habitat' — see Chapter 2). Such burrows can be 20 m long. Snow conducts heat poorly, so the temperature in the burrows is rarely below 0°C despite air temperatures above the snow of −20°C or below. Lemmings build grass nests within the snow burrows and are gregarious, so in large measure they are independent of external temperatures. However, they do rely on their efficient furry insulation when foraging for food, either in the open or in shallow areas of their burrow systems. It has been recognized in recent years that much green vegetation is still available in the winter tundra areas beneath the snow, and this is an important factor in allowing lemmings to breed successfully in both winter and summer (a habit contributing to the enormous increases in population seen in 'lemming years').

Paradoxically, lemmings tend to encounter colder temperatures in their summer burrows than they do in the snow burrows of winter. Lemmings still need to burrow in summer, not least to escape the attentions of the numerous predators (eagles, owls, skuas, wolves, foxes etc.) which rely upon them to raise their young. Because they have to burrow into the earth, they often come close to permafrost layers where temperatures are much lower than at the surface.

Snow burrows or dens are exploited by other arctic endotherms. Rock ptarmigan (*Lagopus mutus*) are amongst the few birds that winter in arctic regions and burrowing into the snow contributes to their survival (also aided by their feathered, widened toes, ideally adapted for walking on snow). Adult polar bears (*Thalarctos maritimus*) normally spend the winter wandering on pack ice in search of food, but spells of poor weather cause male bears to retreat to snow dens, while pregnant and suckling she-bears spend the winter in chambered snow dens on land, often living in them continuously for as much as 6 months without access to food.

One problem of life in burrows for mammals and birds has attracted relatively little attention and certainly merits further study. For burrow life to allow stable body temperatures, ventilation must necessarily be restricted. In consequence because diffusion of gases through soil is fairly slow, CO_2 levels in burrows become quite high, while pO_2 levels may be depressed. Normal atmospheric air contains trace levels of CO_2 and 21% O_2. Measurements in bird and mammal burrows have revealed as much as 6% CO_2 and as little as 14% O_2 (see Boggs *et al.*, 1984 for review). Thus, burrowing endotherms are exposed to combined hypoxia and hypercarbia — both factors which stimulate increased respiratory rate (and therefore work) in non-burrowing birds and mammals. 'Normal' mammals have a respiratory cycle generally driven by blood pCO_2 in the blood supplied to the brain stem where the respiratory centres lie. Raising pCO_2 stimulates ventilation. There is much evidence that burrowing mammals (e.g. echidnas, mole rats, woodchucks) have an elevated 'set point' for such stimulation (Figure 3.10). Haim *et al.* (1985) showed that some fossorial rodents excreted much CO_2 in the form of calcium bicarbonate — apparently to reduce the build-up of burrow CO_2. There is little convincing evidence for an enhanced oxygen-carrying ability in the blood of burrowing mammals, and this is to be expected since the lowest pO_2 values reported from burrows (approx. 110 mm Hg) are no longer than those to be found at an

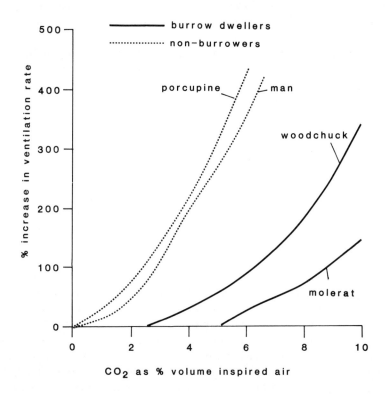

Figure 3.10 Ventilatory responses of mammals to the level of CO_2 in inspired air. Burrowing forms exhibit much less sensitivity to CO_2. Redrawn from Haim *et al.* (1985).

altitude of 3000 m — roughly the level at which high altitude respiratory modifications become necessary. Another problem of life in burrows is connected with sanitation. Many burrowing animals urinate only when outside the burrow, but this is not the case in young animals and birds. Consequently there tends to be a build-up of NH_3 to levels (50–100 ppm) that would be irritating to non-burrowing endotherms. To some extent CO_2 and NH_3 may counter each other, but this is another aspect of burrow existence which merits further investigation.

3.4.2 Nest building

Nest building is exhibited by a variety of animals, but is especially well developed amongst the social insects, mammals and birds. In birds nest building is often substantially decoupled from thermoregulation, especially in warm climates, since it may simply provide a safe and remote place to rear offspring. Similarly, the nests of insects may have many functions beyond protection against thermal extremes. However, in all three groups there are often thermal benefits, and nest building to preserve body temperature is particularly important in mammals.

Two groups of insects are renowned for their nest building abilities, the termites (Isoptera) and the hymenopterans (ants, wasps and bees). All build nests of earth, vegetation (either collected or processed) or secreted material (e.g. beeswax). Termite hills are generally a feature of warm climates so will not be considerd here, but hymenopterans are ubiquitous in temperate areas and quite common in subarctic zones. Honey and bumble bees tend to build their nests in natural holes and burrows, the nest material tending to reduce movements of air, thus slowing loss of heat from these 'social endotherms'.

In wasps the task of thermoregulation is eased by the insulating qualities of the wood pulp 'carton' of the nest. Nest building itself has a regulatory function in these hymenopterans; Potter (1965) reported that low nest temperatures stimulated increased wood pulp foraging in *Vespula vulgaris* so that insulation was rapidly improved.

Most ant species are burrowers and the term 'nest' is a misnomer. However, the red wood ant *Formica polyctena* builds large mounds of vegetation (often of pine needles) which may reach a height of more than 2 metres. It has been suggested by Brandt (1980) that these large heaps are mainly, if not completely, built to achieve thermal stability for the ant colony. Brandt noted that closely related *Formica* spp. do not build mounds, but live in 'nests' (burrow systems) beneath large rocks, which confer thermal stability upon the ant colony by virtue of their high thermal stability. Brandt showed that the loosely built mounds constructed by red wood ants have similar thermal capacities to rock, and that few ants actually live in the mound — most are in soil galleries below. Effectively the mound is a rock replica which enables the species to live in areas without large rocks and which evens out thermal extremes for the underlying community. Brian (1977) reports that daytime temperatures in wood ant nests vary relatively little — ranging between 23 and 30°C.

Most bird nests have relatively little thermal value in themselves because they are cup shaped and do not retain warmth rising from eggs or chicks (though they do provide some shelter from wind). Obviously the situation changes when the parent sits upon the nest, since its body provides warmth (through the brood patches) and a 'lid'. In these circumstances the nest provides a warm environment with a temperature close to that of the adult body. Bird nests are usually made of vegetation and can be very elaborate and thick-walled; often the nests are lined with feathers (e.g. in eider ducks), which improves their insulation considerably, as well as providing a soft, comfortable surface for the young birds to lie on. Eider ducks use considerable quantities of down feathers (hence their value in commercial exploitation) in their nests. This allows the female to leave the nest for short periods, because she covers the eggs with a layer of eiderdown to keep them warm.

A few birds (e.g. the wren, *Troglydytes troglydytes*) build spherical nests (Figure 3.11) which function much like the nests of small mammals (p. 86) and provide enough insulation for the very small hatchlings (5–12 in number) to tolerate brief periods without parental body heat. The importance of nests in thermal terms is particularly difficult to evaluate in the light of the number of Antarctic birds, some of which are quite small (e.g. terns, cape petrels) yet nest on the open ground, their nests consisting of a few small stones or a scrape in the ground.

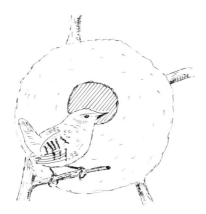

Figure 3.11 Drawing of wren (*Troglodytes troglodytes*) and its spherical nest. Male wrens build the outer framework of several nests at the onset of the breeding season; his mate eventually selects one of these and lines it with fine material.

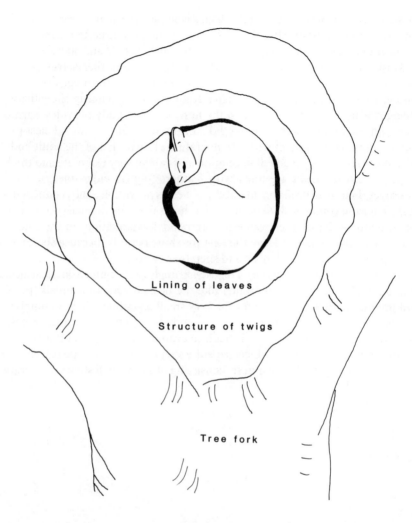

Figure 3.12 Nest (or drey) of red squirrel (*Sciurus vulgaris*). Note the thickness of the insulating layers, the tightness of the central cavity around the animal and the curled posture of the squirrel — all features which result in high nest temperatures.

Most study of the thermal value of nests has been provided by studies upon small mammals. Many mammals build nests within burrows, and thus effectively surround themselves with an extra fur coat, others build nests above ground, either amongst vegetation (e.g. harvest mice), or in tree holes. A few occupy the nests of birds; European dormice sometimes take over the nests of wrens. Generally speaking, mammalian nests are spherical with one or two entrance holes. Almost any convenient nesting material will be used (grass, moss, leaves, plastic, paper or even aluminium foil). However, there is a tendency towards the use of coarser material for the exterior of the nest and finer material (including moulted fur) for the lining. Pearson (1960) found that the nest of the harvest mouse (*Reithrodontomys megalotis*) allowed a single occupant to reduce heat output by 20%. When combined with gregariousness and/or torpor, nests can allow savings in metabolic rate of well over 50% (e.g. Casey, 1981; Vogt and Lynch, 1982). Similar data have been presented for various tree squirrels which overwinter in nests or dreys (but do not hibernate as they are popularly assumed to do). Havera (1979) studied arboreal fox squirrels (*Sciurus niger*) and found that these animals did not need to increase their metabolic rate when inside nests until the external temperature fell below −8°C. Air temperatures inside squirrel nests are much higher than outside, pointing to the efficient drey design (Figure 3.12) which involves thick layers of twigs, lined with leaves, feathers, fur, moss and lichens (Laidler, 1980). Pullainen (1973) found that temperatures within red squirrel (*Sciurus vulgaris*) nests in Finland were 20–30 deg. C higher than an external temperature of −5°C, again pointing to considerable energy savings. Whilst within nests, many small subarctic mammals are creating warm temperate microclimates around them.

4 Anatomy and physiology of endotherms

4.1 SHAPE, SIZE AND CLIMATE

The rate of heat loss (whether by radiation or evaporation) from a warm body to a cool environment is proportional to the surface area of the body; the lower the surface area-to-volume ratio, the slower the loss of heat. A number of 'Zoogeographical Rules' have been suggested from this and other physical laws.

4.1.1 Bergmann's Rule (1847)

This rule states, for endothermic species, that amongst related animals, those found in colder areas are larger than those found in warm areas. Mayr (1956, 1963) has stressed that a zoogeographical 'rule' is only valid if it is true in more than 50% of cases and suggests that Bergmann's Rule should only be applied strictly to within-species variation in size. However, such rigour has rarely been seen in the textbooks which have disseminated the idea that size increases towards the pole. It is relatively easy to identify examples which appear to support the validity of Bergmann's Rule. Polar and Kodiak Brown bears are considerably larger than tropical bears; the largest penguins occur on the Antarctic mainland, the smallest in the tropics. Siberian tigers are larger than their Indian cousins. Amongst ospreys (*Pandion haliaetus*), widely distributed in temperate and subtropical areas, there is a clear tendency for size to increase towards the north (Prevost, 1983) (Figure 4.1). Bergmann's Rule has also been shown to work well for changes in altitude, with larger birds and mammals being found at greater (= colder) heights above sea level. However, there are many exceptions to the Rule. Tiny snow buntings and lemmings thrive in the arctic, while the largest of living birds and terrestrial mammals (ostriches and

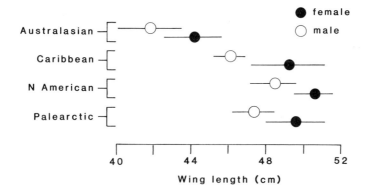

Figure 4.1 Size of ospreys (*Pandion haliaetus*) from different latitudes; an expression of Bergmann's Rule. High latitude ospreys are substantially bigger than those living in subtropical areas. Redrawn from Prevost (1983).

elephants respectively) live in the tropics. Bergmann's Rule also fails completely in the case of burrowing mammals including rodents and insectivores such as the European Mole (*Talpa europaea*). Even in those cases where larger animals within a species do seem to live in colder latitudes, it may be argued that larger individuals will be selected for because they will be capable of storing larger quantities of energy to withstand the longer periods of adverse weather conditions which are characteristic of high latitudes, and not because of crude surface-area to volume considerations. Given the remarkable controllable insulation of fur and feathers, the efficiency of peripheral vasomotor control of heat loss (and the accompanying evolution of countercurrent heat exchangers [q.v.]), it is naive to regard an endotherm as a simple warm, physical body exposed to a cold environment. The author believes that Bergmann's Rule has had value in the past as a stimulus to thought and experiment, but is now much exposed by its exceptions, so needs to be used with great care.

4.1.2. Allen's Rule (1877)

Allen's Rule (which in essence is an extension of Bergmann's Rule) states that 'in warm-blooded species, the relative size of exposed portions of the body decreases with decrease of mean temperature'. Extremities (tails, ears and beaks) tend to be smaller/shorter in a cool environment, thereby reducing surface area and heat loss. This rule seems to be more convincing than Bergmann's Rule, particularly in the case of mammals. High latitude rodents have relatively smaller ears, tails and digits than do their relatives from the tropics and subtropics. A particularly striking example is that of living and fossil elephants. Thanks to the discoveries of mummified woolly mammoths (*Mammuthus primigenius*) over the past 200 years in northern Russia, it has been established that these tundra-dwelling elephants had tiny ears by comparison with their tropical relatives, particularly the African elephant which uses its huge ears to release body heat. Polar bears have smaller ears than do tropical bears, while races of dogs, deer and horses from colder climates tend to have shorter limbs. Again, though, there are exceptions. Amongst humans (Chapter 8) there seems to be a relatively weak correlation between size of extremities and climate, while polar birds, being predominantly aquatic, have large webbed feet. The large, globular chick of the albatross may seem the perfect expression of Allen's Rule, yet the slim-winged, streamlined adult flying machine derived from it is quite the opposite!

4.1.3 Gloger's Rule (1833)

Gloger's Rule relates surface colour to temperature and humidity. Gloger postulated that dark colour due to melanin pigmentation is less developed at high (cold) latitudes. Much of the relevant literature is devoted to humans (Chapter 8) and the effects of ultraviolet radiation, but for other animals this Rule seems rather dubious. Whether ectotherms or endotherms are considered, it is possible to identify light and dark coloured species at all latitudes. The physics of heat absorbance and reflection are not simple; the requirements of cryptic camouflage also complicate matters.

4.1.4 Mayr's rules

Mayr (1963) presents a number of zoogeographical rules for birds. At high latitudes in the north wings and tails tend to be longer, egg clutches to be larger and migratory behaviour to be better developed. It has also been found that the digestive and absorptive parts of the gut are relatively larger in more northerly bird populations.

4.2 STRUCTURAL INSULATION: FUR, FAT AND FEATHERS

All mammals have hairy skins; all birds possess feathers. Both groups commonly deposit lipids in their subcutaneous tissues. The insulative function of these three features is now so generally accepted, that it is easy to forget that little quantitative information about their value existed until a group of scientists led by Per Scholander carried out a series of elegant studies in the late 1940s upon arctic and tropical mammals and birds (Scholander *et al.*, 1950a, 1950b, 1950c). They found that subcutaneous fat is a bulky and heavy insulator, of little importance to arctic terrestrial mammals (though of considerable importance to seals, walruses, polar bears and the smaller whales). Arctic

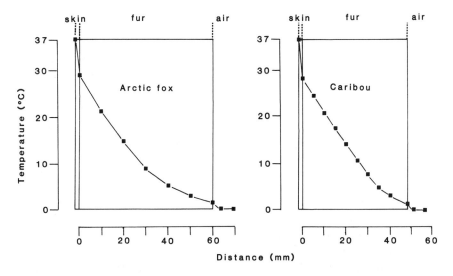

Figure 4.2 Thermal gradients set up in the fur of white fox (*Alopex lagopus*) and caribou (*Rangifer tarandus*). These data were derived experimentally on dead skins held on a warm metal plate (plate temperature 37°C) and exposed to a cold atmosphere (0°C). Redrawn from Scholander *et al.* (1950a).

land mammals depend upon their fur for insulation and, in large animals, this is remarkably effective since a complete thermal gradient between body temperature and ambient temperature can occur within the fur (Figure 4.2).

4.2.1 Fur

Fur is an insulator because its presence causes a layer of still air to be held against the body. Air is a poor conductor of heat and has a relatively low heat capacity; the keratin of hair is also a poor conductor and the small cross-sectional area of hairs means that little heat is carried to the environment by the hairs themselves.

The larger arctic mammals (marten, rabbit, husky dogs, foxes, beavers and bears) all have fur thicknesses of between 30 and 70 mm. In contrast, tropical mammals rarely have fur more than 20 mm thick, only the tree sloths slightly exceeding 30 mm. Generally, large arctic mammals have fine, closely set fur which traps essentially immobile air. The polar bear, *Thalarctos maritimus*, which spends much of its life in water is something of an exception to this pattern as the fur is coarse. However, the polar bear inhabits pack ice when foraging for its seal prey and therefore regularly enters water of freezing temperature. It seems that the bear's coarse pelt enables it to shake water out of its fur quickly and effectively when it climbs out of water. The large size of the polar bear (and consequent low surface-area to volume ratio), the length of the fur (70 mm) and the hollow nature of the individual hairs offset the disadvantages of pelt coarseness to some extent. The hollow hairs reflect visible light (hence the white appearance of the pelt), but are transparent to ultraviolet wavelengths. The hairs act as light guides to transmit the ultraviolet light from the tips of the hairs to the skin, thus heating it. A similar arrangement has arisen by convergent evolution in the white Rocky mountain goat (*Oreamnos americanus*). Interestingly, for arctic mammals of dog size and above (5–500 kg), there is no significant correlation between degree of insulation and body size; their fur appears to be of maximum useful length.

Scholander *et al.* (1950a) also demonstrated that fur is a remarkably poor insulator in water, especially if the fur is completely wetted as is the case for most pinnipeds. The coarse-furred polar bear, which cannot trap air, loses heat at 20–25 times the aerial rate in still ice water; under the more realistic conditions of agitated ice water heat loss was dramatically increased to 45–50 times the rate in air. A later study by Frisch *et al.* (1974) showed that heat loss would be even greater in polar bears (and in the harp seal *Pagophilus groenlandicus*) were it not for the existence of a woolly underfur (10 mm thick in the case of the polar bear) which retained a layer of stagnant water against the skin.

A variety of semi-aquatic mammals, including the northern fur seal (*Callorhinus ursinus*), beaver (*Castor canadensis*), the mink (*Mustella vison*), muskrat (*Ondatara zibethicus*) and otters (*Lutra* and *Enhydra*) have fine fur which does trap an insulating layer of air. However, even the beaver, which has such a fine, closely packed pelt that a layer of air several mm thick is trapped against the skin during immersion, still loses heat 10 times as fast in still ice water as it does in air of the same temperature (0°C). Perhaps surprisingly, the duck-billed platypus (*Ornithorhynchus anatinus*) has the most effective fur insulation in cold water (Dawson, 1983). Platypuses live in the highlands of eastern Australia and forage for many hours in water close to 0°C despite a body weight of only 1 kg. Contrary to popular opinion, the monotreme *Ornithorhynchus* has an efficient metabolism and well developed countercurrent heat exchangers at the roots of the limbs and tail; a core temperature of about 32°C is maintained. The fur is extremely tightly packed (Dawson compared it with the fur of the sea otter *Enhydra lutris*). The underfur fibres are kinked and the guard hairs are flattened, so that a layer of air is trapped against the skin throughout the time in water. The net result is that the platypus only loses heat at 2–3 times the rate in air, but this probably represents the greatest attainable efficiency of fur as an insulator in water.

Kruuk and Balharry (1990) have recently noted that, while a large number of semiaquatic mammal species enter fresh water, only two (mink and otter, *Lutra lutra*) enter sea water on a regular basis. In studies of otter pelts they have shown that small salt crystals stick the hairs together when the pelt dries during periods when the otters are out of water. This degrades the insulating properties of the fur, not only in air, but also when the animal returns to water. Kruuk and Balharry demonstrated that *Lutra lutra* will not readily enter sea water unless it has access to fresh water as well to wash away the salt. This limits the area of coast which may be exploited by 'land' otters. They also noted that otters which had been swimming in sea water spent much time in grooming and pelt maintenance. Studies of sea otters (*Enhydra lutra*), the only fully marine mammal to rely on fur rather than blubber for insulation, reveal that it spends a great deal of its life (several hours per day) in grooming the fur (Shimek and Monk, 1977), presumably to remove salt crystals in the part of the pelt exposed to the atmosphere, and also to restore the ordered nature of the fur in an analogous fashion to the preening of bird feathers. Baby sea otters possess so much air in their pelt that they are incapable of diving; their mothers spend time in grooming their coats too — an ecological cost which must severely limit foraging time.

Small arctic mammals (lemmings, squirrels, weasels and shrews) necessarily have shorter and lighter fur or their locomotion would be impaired. In fact the thickness of their fur is similar to that of tropical mammals, and they depend to a large extent on life in nests and burrows (often communal) to minimize the periods of exposure to the full rigours of a high latitude environment. However, their fur is denser than in tropical mammals of similar sizes, and their limbs and head are covered by more extensive fur than their temperate or tropical relatives. Lemmings provide good examples of a small, furry arctic mammal. Short-eared, short-limbed and effectively tailless they have dense fur covering the whole of the body and head. The skin is loose, allowing the insulative value off their fur to be preserved as they squeeze through earth and snow burrows.

Fur thickness is controllable to a certain extent as the angle of each hair to the skin can be altered by muscle action. Full contraction of the piloarrector muscles (Figure 4.3) brings each hair to a position perpendicular to the epidermis ('piloerection'). However, the relationship between hair angle and the insulative value of fur is not simple. It is possible that partial erection of previously flattened fur will improve insulation, but complete erection of the hair is more likely to increase heat loss by providing routes for convective heat loss. Piloerection in different parts of the body is likely to have different effects. Since warm air rises, complete piloerection on the dorsal surfaces may cause heat loss, while piloerection of the belly fur may conserve heat.

Before leaving the topic of fur it needs to be stressed that the quantity and thickness of fur in mammals may vary during their life cycle or with the season. Seals, for example, have thin sparse hair as adults, but seal pups have a temporary, thick, fine pelt during the first few weeks of life when they are effectively terrestrial animals, incapable of swimming. It has been noted that the pups are deserted by their mothers during bad weather as the latter cannot tolerate air temperatures below $-20°C$ and have to retreat to the sea to avoid

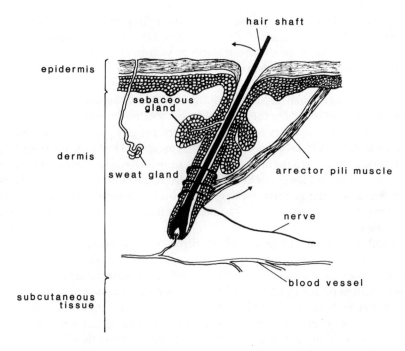

Figure 4.3 Semi-diagrammatic section of mammalian skin. Contraction of the erector pili muscles elevate hairs to make the pelt effectively thicker; the action of these muscles is antagonized by the elasticity of the dermis.

hypothermia (Pierotti and Pierotti, 1983). A large number of mammals living at high latitudes demonstrate a rather less dramatic alteration in pelt thickness as they grow and shed their 'winter coats'. Such seasonal changes have been investigated particularly in the arctic fox (e.g. Hammel, 1955; Underwood and Reynolds, 1980), but there is also good evidence that the woolly mammoths (*Mammuthus primigenius*) of the Pleistocene Siberian tundra shed their long reddish fur in the summer (see Sutcliffe (1985) for discussion). It is also probable that fur thickness can be 'fine tuned' to some extent to meet the challenges of particularly severe or mild winters, since it has long been known that some mammals can respond to prolonged cold by increasing their fur length (e.g. dogs — Hoesslin, 1888; rabbits — Mayer and Nichita, 1929). Such phenotypic morphological responses to cold merit further investigation.

4.2.2 Feathers

The feathers of birds have a similar insulative function to fur, but their much greater structural complexity and muscular controllability have several implications. First, from an experimental point of view it is impossible to duplicate the ordered plumage of a living bird in estimates of insulative value performed on dead skins. Scholander *et al.* (1950a) appreciated this, but nevertheless attempted estimates on the skins of snow buntings and ptarmigan. These isolated skins, on which the feathers were probably flatter than in life, had insulative values close to those of the fur of mammals of comparable size (lemmings and marten respectively), which tends to suggest that the plumage of live birds is probably more effective than fur.

Second, birds can increase the thickness and insulative value of their plumage considerably ('fluffing up') and such behaviour is seen in resting and roosting birds in cold climates. However, as with fur, maximal erection of the plumage probably serves to increase heat loss rather than reduce it.

Third, the complex orientation and structure of feathers is particularly important to insulation in aquatic birds. Birds have thin, dry, loose skins which possess neither sweat nor sebaceous glands. Down feathers, next to the skin, are soft with many processes; contour and flight feathers feature multiple subdivision of elements (rachis, barb, barbule; Figure 4.4). Consequently, the plumage presents a thick, ultra-fine, many-layered mesh to the environment. All birds possess a single cutaneous gland, the uropygial or preen gland, sited at the base of the tail. Oil from the preen gland is used to clean and lubricate the feathers, but is particularly copious in aquatic birds, especially those from cold climates. The combination of water-proofing preen oil and feathery mesh allows birds to trap air next to the skin when in water, thus achieving a most effective insulation which is instrumental in allowing large numbers of relatively small endotherms (e.g. auks, petrels) to exploit the high productivity of cold seas (all high latitude marine mammals are fairly large). The air-filled plumage of aquatic birds does gradually leak as preen oil is lost and long periods are consumed in the maintenance of good feather order by preening whilst resting on land. It is noticeable that aquatic birds spend far more time in preening (< 2–3 hours per day) than do terrestrial species, and this extra time when foraging is impossible represents an ecological 'cost'. The cost may be offset to some extent by the ingestion of sun-irradiated preen oil — which contains vitamin D!

The trapped air within plumage also has an additional function which bears upon insulation — it creates considerable positive buoyancy. This buoyancy, coupled with the general lightness of bird body architecture, aids diving birds in their rapid return to the water surface. However, diving occupies a relatively small proportion of the life of most aquatic birds. Far more time is spent resting or swimming at the sea surface and the buoyancy of the plumage results in a large proportion of the animal projecting above that surface. This situation is unique to birds as aquatic mammals are generally at least 80% immersed when swimming at the surface (usually $>90\%$). The effects of simultaneous exposure to both air and water are undoubtedly complex. At one extreme, an

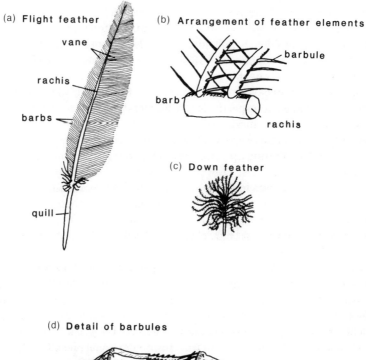

(a) **Flight feather**

vane

rachis

barbs

quill

(b) **Arrangement of feather elements**

barbule

barb

rachis

(c) **Down feather**

(d) **Detail of barbules**

distal barbule

barb   hamuli (hooks)

proximal barbule
(quill side of barb)

Figure 4.4 Structure of feathers. Note that most insulation is provided by down feathers, but integrity of insulation requires regular preening of contour feathers (not shown but similar in general arrangement to flight feathers) and flight feathers as well as the down. Preening ensures that the hamuli of distal barbules are 'zipped' onto the ridges of proximal barbules to maintain an ordered meshwork.

aquatic bird swimming in cold polar water on a still spring or summer day may lose less heat because much of its body is above the water line. Conversely, under conditions of subzero air temperatures and high winds the exposure of the head and dorsal surfaces may be disadvantageous because of wind-chill. However, aquatic birds usually retreat to terrestrial shelter during poor weather, and tend to migrate to lower latitudes in winter anyway, so on balance it is probable that the high floating posture is of insulative advantage.

For one group of birds, the penguins, diving does occupy a large proportion of their life. When they are swimming near the sea surface, their thermal behaviour is little different from that of other sea birds. However, some penguins are deep divers; the emperor penguin (*Aptenodytes forsteri*) dives to at least 265 metres (probably in search of squid). The problem with dives to such depths is that the birds are exposed to high pressures (roughly 26.5 atmospheres in this example) which considerably reduce the volume of the air in the plumage. This has two effects. First the insulation of the plumage is lost, causing rapid heat loss. Second, the ordered nature of the plumage is disturbed. Emperor penguins respond to the heat loss by muscular thermogenesis (shivering); they restore the order of the plumage by combing it with the bill, together with preening.

4.2.3 Fat

Fat, as a material, seems at first consideration to be a useful insulating material. It has a low thermal conductivity and, as far as biological materials are concerned, has a relatively low density. However, considerable caution must be exercised when evaluating the function of lipid deposits in a particular animal. First, fat can act as an effective insulator only if it is superficially distributed and generally poorly supplied with blood. If a fat deposit is well vascularized it can only function as an insulator when neurohumoral influences cause peripheral vasoconstriction otherwise the warm blood flowing through fatty tissues delivers heat to the body surface.

Second, it should be recognized that fat can have several other functions beyond that of insulation. Primarily, fat acts as an energy reserve and large deposits are a common feature of large mammals (whether polar or tropical, terrestrial or aquatic) in which it helps to give independence of day-to-day fluctuations in food availability (small mammals cannot afford the energetic 'cost' of carrying large fat reserves around). Thus, although great whales (e.g. the blue whale, *Balaenoptera musculus*) have enormous quantities of peripheral blubber, it should not be assumed that these are mainly for insulation; the very low surface-area to volume ratio of such extremely large animals, plus efficient vasomotor control probably make such thick insulation unnecessary, at least in tropical and temperate waters. Instead, the blubber fuels long breeding migrations and periods of the animal's life history when krill swarms are unavailable or sparse. Indeed, it seems probable that the larger whale species have difficulty in keeping their body temperatures down rather than up (Parry, 1949; Kanwisher and Sundnes, 1966; Ridgeway, 1972); in the early days of whaling it was noticed that dead whales (in which circulation was absent and water flow over the body surfaces negligible) showed rapid internal 'cooking' (within a few hours of death) if they were not gutted and cut up, presumably because of undispersed metabolic heat.

Fat also provides buoyancy in aquatic animals since it is less dense than water, offsetting the high densities of bone and muscle. Lipid deposits may also contribute to hydrodynamic stability; Lockyer *et al.* (1984) noted that fin whales (*Balaenoptera physalus*) had much thicker dorsal than ventral blubber; the dorsal placement of low density material is likely to resist any rolling tendency.

In deep diving mammals (seals, toothed whales), which spend long periods under conditions of high ambient pressure, fat deposits provide a nitrogen sink (nitrogen is more soluble in fat than in water) which prevents nitrogen bubbling out of solution when the animal decompresses as it surfaces, thereby avoiding decompression sickness (the 'bends').

Finally, fat commonly acts as a 'shaping' structural material. In humans the sexual signalling significance of female fat deposits seems established beyond doubt, while in aquatic mammals and some birds (e.g. penguins, blue-eyed cormorants (*Phalacrocorax atriceps*)), blubber contributes significantly to their streamlined, drag-cheating shape. Lockyer *et al.* (1984) provided evidence that the collagenous blubber acted as an exoskeleton in fin whales, functioning as a hydrodynamic buffer and perhaps promoting maintenance of laminar flow over the animals during fast swimming.

Obviously none of these functions of fat exclude an insulative function (there is evidence, for example, that the subcutaneous fat deposits of women contribute to a greater tolerance of exposure to cold than is shown by men), but the author simply wishes to stress that the proposition that peripheral fat is an insulative material must always be treated with great caution.

'Fat' is a portmanteau term describing many sorts of triglyceride mixtures. From the earliest years of this century, scientists have progressively revealed relationships between the physical/biochemical characteristics of depot fats of mammals and the features of their thermal environments. It has long been known that the tissues of poorly insulated extremities of resting arctic birds and mammals are regulated to much lower temperatures than the core body temperature (35–40°C). Irving and Krog (1955) showed that the tissues of hooves, foot and toe pads, feet of birds and the noses of arctic mammals are often close to 0°C — nearly 40 deg. C below the core temperature and perhaps 30 deg. C below the temperature of skin beneath fur or feathers. Many years earlier Henriques and Hansen (1901) had demonstrated that the melting point of the subcutaneous fat of pigs depended on their previous thermal history. The fat of pigs maintained at 30–35°C melted at a temperature 1.4 deg. C higher than that of pigs held for long periods at 0°C. These ingenious workers even kept a pig at 30–35°C, but with a sheepskin overcoat permanently in place; the melting point of this animal's fat was 2.4 deg. C higher than an undressed pig! A number of studies in the 1940s and 1950s built on this observation. First, Schmidt-Nielsen and Espeli (1941) noted that marrow fat taken from the femur (proximal hindlimb bone) of an ox and a reindeer (*Rangifer tarandus*) melted at about 40°C, yet fat from their feet was fluid when cold. K. Schmidt-Nielsen (1946) discovered that melting points of fats taken from the legs of several humans declined from thigh to foot. There was also cultural evidence from Inuit communities that caribou fats taken from different parts of the body had different usages which were related to fluidity. For example; fat used as gun oil could only be obtained from terminal limb bones. All of these observations culminated in a classic study by Irving *et al.* (1957). These workers demonstrated for a range of medium sized and large terrestrial arctic mammals (red fox, *Vulpes vulpes*; wolf, *Canis lupus*; sledge dog, *Canis familiaris*; reindeer/caribou, *Rangifer tarandus*) that there are gradients of fat melting point in the limbs, and that the limb fats melt at lower temperatures than fat taken from central deposits. Reindeer nose fats also have low melting points. Gradients of melting point were much more dramatic in the long-limbed caribou than in dogs/foxes. Irving *et al.* also used earlier data (Irving and Krog, 1955) to show that pronounced decreases in fat melting point occurred where the legs of caribou became slender and relatively hairless; the more fluid fats were found distal to vascular countercurrent heat exchangers. The next contribution of the study was to show that low fat melting points were associated with greater proportions of unsaturated fatty acids in the lipid deposits. This study has since been widely quoted as an example of arctic adaptation, but Irving *et al.* made no such claim, since they also found gradients of fat melting point and degree of fatty acid saturation in legs of the subtropical Panamanian brocket deer, *Mazama americana*. They also knew that such gradients existed in humans (Schmidt-Nielsen, 1946) which are essentially tropical animals (Chapter 8). However, it does seem that arctic terrestrial mammals do exhibit a more extreme version of a well-established mammalian trend. Irving *et al.* suggested that more fluid peripheral fats could allow greater mobility of appendages. This may be so, but the author wishes to venture an alternative, more generalized hypothesis. All terrestrial mammals and birds encounter environmental temperatures below the core temperature; even in the tropics, nights are sometimes cold. All endotherms must therefore have peripheral tissues which are functional at subcore temperatures for prolonged periods, since it would be energetically inappropriate to keep the whole body at core temperature. All endotherms appear to have efficient neurohumoral control over the peripheral circulation which allows them to reduce the flow of warm blood to the extremities. To be functional, tissues have to be biochemically active — membranes must be fluid (Chapter 1) and biochemical substrates within cells must be accessible. This means that storage lipids must be in sufficiently fluid state for exchange to take place. In addition, it must be realized that no endotherm, whether arctic or tropical, can allow its peripheral tissues to fall much below 0°C if damage is not to occur. This means that there is relatively little difference between the foot temperature of a reindeer living at −25°C, and the foot temperature of a camel exposed to near-freezing conditions on an Arabian desert night. Arctic mammals and birds require much more efficient insulation by fur and feathers than do their tropical relatives, but the amount of special tissue adaptation needed in the extremities is surprisingly small.

Table 4.1 Freezing points and fatty acid composition of neutral lipids extracted from the adipose tissues of various animals

Species/tissue	freezing point (°C)	% saturated fatty acids	% mono-unsaturated fatty acids	% poly-unsaturated fatty acids
Leatherback turtle				
carapace blubber	+17.0	42.6	40.2	17.2
flipper adipose	+11.4	37.9	48.2	14.1
Grey seal blubber	−4.0	16.8	55.4	27.8
Sheep subcutaneous fat	+42.5	53.1	44.9	2.0

Note that the leatherback turtle probably has a core temperature of around 25°C, but has countercurrent heat exchangers at the roots of the flippers (which therefore operate at near-ambient temperatures)
After Davenport *et al.*, 1990b.

Fluidity of fats has been correlated with their fatty acid composition; Irving *et al.* demonstrated that low melting points are associated with increased concentrations of unsaturated fatty acids. Recently the author (Davenport *et al.*, 1990b) has studied lipids of the leatherback turtle *Dermochelys coriacea* which appears to be endothermic when foraging in cold temperate waters, but presumably maintains its peripheral tissues at low temperature. The blubber of this animal contains a greater quantity of monounsaturated and polyunsaturated fatty acids than do the fats of terrestrial mammals in which skin temperatures are high and insulation provided by fur (e.g. sheep — Table 4.1), but an even greater level of unsaturation is found in the blubber of the grey seal (*Halichoerus grypus*) which is fluid down to −4°C.

The relationship between fat distribution/composition and evolution is a complex one. High latitudes are associated with alterations in food availability (because of changes in photoperiod and light intensity) as well as low temperature. Adipose stores of energy are necessary to smooth out fluctuations in food availability, and, for straightforward geometric reasons, it is easier to accommodate bulky material in thin peripheral layers rather than centrally. These factors provide preadaptation for the use of fat as an insulating material. For marine endotherms, fat is much more readily available at high latitudes, because the ectothermic animals at the base of the food webs (copepods, euphausids) accumulate lipid to overwinter. Terrestrial carnivores gain similar advantage by eating lipid-rich prey. Much of the unsaturated fatty acid component of endotherms is therefore acquired directly from the food chain, and does not require synthesis *de novo* (see Sargent *et al.*, 1987 for discussion).

4.3 VASCULAR ARRANGEMENTS TO MINIMIZE HEAT LOSS

Great plasticity of blood vessel arrangements is characteristic of vertebrates, but a general description of the blood supply and drainage of an organ would be as follows. A small artery, leading from the main systemic circulation breaks up into a number of smaller arterioles. These supply blood to a network of fine capillaries, where all of the exchange functions of blood (movement of gases, heat, nutrients, hormones etc.) take place. The network of capillaries delivers blood to venules and thence to veins which are closely associated with the arteries supplying the organ and deliver blood back into the systemic circulation. In endotherms, particularly those living in cold environments, this basic pattern has been modified many times (by convergent evolution) to restrict heat loss in appendages (flipper, legs, tails). Firstly, there is a tendency towards the evolution of arterio-venous anastamoses, short-circuiting 'shunts' which are relatively wide-bore vessels which join small arteries and veins. Under cold conditions warm blood flows through these shunts (which are situated in the core of extremities) rather than moving through more peripheral capillary beds. Direction of blood flow through the shunts is under neurohumoral control, basically mediated by the hypothalamus, and acting through the muscles regulating the bore of the vessels involved. Arteriovenous shunts may also be dilated widely for short periods to allow the rapid flow of warm blood into extremities to avoid low temperature tissue damage.

An even more specialized vascular arrangement is seen at the roots of tails, feet, legs and flippers. This is the 'countercurrent heat exchanger' (Figure 4.5) or *rete mirabile* ('marvellous network'). Such arrangements involve the arterial supply to an extremity breaking up into a large number of short arterioles which then rejoin to form larger vessel(s). The venous drainage of the extremity breaks up into a similar number of small veins which are closely applied to the arterioles, before joining up to form larger veins. The effect is to expose a large surface area of thin tissue containing warm blood, to a similar area containing cold blood. Heat flows down the temperature gradient, so that, within a short length of extremity, most of the heat in the outgoing arterial blood is transferred to the incoming venous blood. The 'price' of this arrangement is that the extremity is cold and supplied with cool blood. In turn, this means that the tissues of the extremities must be functional at temperatures well below the core body temperature. Particularly important is the maintenance of membrane fluidity in cells and their organelles. Enzyme function must also be regulated at these temperatures (e.g. Malan, 1979). The complexity of countercurrent heat exchangers is variable. Mammals with relatively long, well insulated legs, may not have a *rete mirabile*. Øritsland (1970) reported that the heat exchangers of the leg of the polar bear (*Thalarctos maritimus*) consisted of lengths of artery and vein which were closely applied over lengths of 30–60 cm.

(a) Countercurrent principle

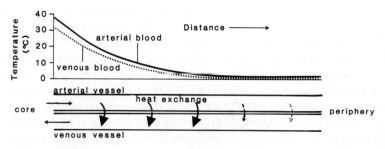

(b) Rete mirabile

(c) Arrangement of leg blood vessels in high latitude birds

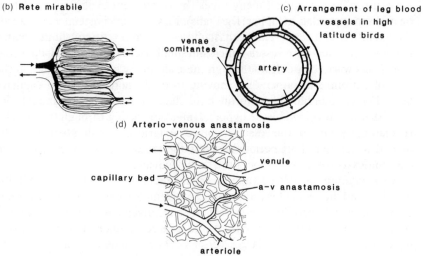

(d) Arterio-venous anastamosis

(e) Diagram of blood flow arrangement in appendage

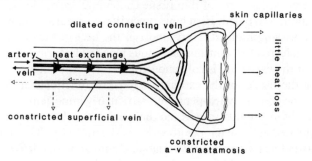

Figure 4.5 Countercurrent heat exchange systems and vascular arrangements to minimize heat loss (see text for details).

In this case the transfer of heat from arteries to veins must take place relatively slowly. Mitgård (1980, 1981, 1989) has paid particular attention to countercurrent heat exchange in the legs of birds. He has demonstrated a third type of heat exchange system, particularly well developed in arctic seabirds such as the Iceland gull, *Larus glaucoides*. In these birds the venous return from the feet takes the form of 3–4 veins (venae comitantes) intimately associated with the single artery. As in polar bears this results in heat exchange taking place over a relatively long distance. Mitgård also showed that the area of contact between veins and artery was greater in the Iceland gull (which breeds in Greenland) than in the closely related herring gull (*Larus argentatus*) which has a more temperate distribution.

4.4 PHYSIOLOGICAL INSULATION

All mammals and birds possess temperature gradients within their body tissues. Even at rest within the thermoneutral zone the liver (a thermogenic organ) of a dog is about 1 deg. C warmer than arterial blood, while the temperature of the blood flowing through the right side of the heart is usually about 0.2 deg. C above that in the left side (heat being lost in the lungs). The mouth and nasal passages of endotherms are cooler than the core, while the gut (particularly in herbivorous animals with fermenting chambers in the large intestine) may be a degree or two warmer.

Endotherms can control the distribution of blood flow to tissues by virtue of vasoconstriction and vasodilation, the narrowing and widening of small arteries and arterioles achieved by contraction and relaxation of smooth muscle in their walls. Because the fluid flow for a given pressure difference through a pipe is proportional to r^4 where r is the radius of the pipe, slight changes in blood vessel diameter can have great influence on blood flow, though blood vessels do not behave quite like rigid pipes! By controlling the flow of warm blood to peripheral tissues, an endotherm can alter the temperature gradient between the 'core' (brain, heart, liver, lungs) and the surface of the skin, and thus change the rate of heat loss. In cold environments relatively little blood is supplied to the periphery.

Blood flow to a given tissue is controlled in two ways. First, there is centralized feedback reflex control of blood distribution by the hypothalamus (Figure 4.6). Classic experiments on dogs showed that the hypothalamus is itself temperature sensitive (reacting to changes in the temperature of blood supplied to it), with the anterior part of the hypothalamus containing a 'thermostat' and the posterior part being cold sensitive. However, there are other central thermoreceptors (e.g. in the spinal cord), while the hypothalamus is also supplied with information by peripheral temperature receptors in the skin and oesophagus. In passing, it should be noted that the peripheral cold

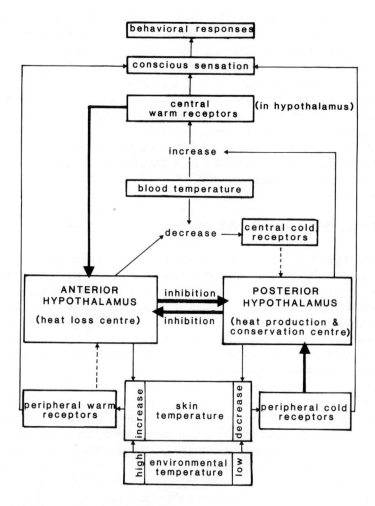

Figure 4.6 Diagram to show the possible relationships between blood and environmental temperatures and the hypothalamic centres. Probable sources of conscious thermal sensation (and hence behavioural responses) are indicated. Those pathways which are thought to be most important are indicated by bold solid lines, pathways of lesser importance by fine solid lines, pathways of debateable significance by dashed lines. Slightly modified from Hardy (1976).

receptors probably also supply information to the cortex of the brain, resulting in conscious sensation of cold — which may trigger behavioural responses (e.g. huddling, burrowing etc.). The hypothalamus 'set points' (i.e. temperatures at which heat dissipation or heat conservation are triggered) are stable in normal, active animals, but are changed by alterations in some plasma ion concentrations (particularly Ca^{++}, and Na^{+}) or the presence of bacterial pyrogens (fever-inducing substances). The set points are also altered during torpor or hibernation, though the factors which induce the changes are unclear. The hypothalamus controls blood flow via the brain stem vasomotor centre and sympathetic nervous system. Vasoconstrictor nerves supply the smooth muscle of the walls of small arteries and arterioles; when they fire they release noradrenaline which excites the smooth muscle and causes contraction and reduced blood flow. Given the huge number of nerves and blood vessels involved, it is obvious that considerable subtlety of perfusion control is possible, so that different areas of peripheral tissue are held at different temperatures. Relatively inactive tissues (e.g. blubber, skin of limbs) tend to be held at lower temperatures than areas of the scalp and face. Commonly the belly of mammals is maintained at a higher temperature than the dorsum — because heat rises. It appears that the normal state of the blood supply to a given tissue involves a degree of vasoconstriction ('vasoconstriction tone'). If sympathetic output (i.e. firing rate in vasoconstrictor nerves) is reduced as a result of rising core body temperature, the blood vessels dilate passively as a result of the pressure difference between the lumen of the blood vessel and the tissues surrounding it. This dilatation can increase the flow of blood to peripheral tissues by two orders of magnitude (i.e. more than a hundred-fold). It can be an important response in well insulated high latitude endotherms, particularly under conditions of exercise combined with warm weather when overheating is a serious risk.

Second, there is local control of perfusion by the concentration of metabolites in the tissues. This is the least understood of control mechanisms, but protects the tissues against damage. If the blood supply to a particular tissue is interrupted or reduced by vasoconstriction, there is a loss of oxygen and energy, plus a build-up of metabolites within that tissue. The build-up of metabolites eventually causes the blood vessels supplying the tissue to dilate ('cold vasodilatation') and allow blood to flow again ('reactive hyperaemia'). Once the metabolites have dispersed, centrally-mediated vasoconstriction is re-established and blood flow reduced again. A cycle of vasonstriction and cold vasodilatation will persist throughout the period of exposure to cold. Several metabolic factors could be responsible for inducing the vasodilatation (e.g. falling pO_2, rising pCO_2, falling pH, increasing tissue osmolarity (because of accumulation of small, osmotically active molecules)), and the reaction could be produced by 'local hormones' such as the peptide bradykinin. Most of our knowledge of local control of perfusion is based upon experiments on a limited range of mammals (particularly cats, dogs and humans), and the importance of cold dilatation in marine mammals or high latitude terrestrial mammals is essentially unknown. Bird plumage is usually so efficient in insulating the animal, that skin temperatures are high and the superficial tissues richly perfused. However, feet and wings often function at low temperatures, so will show the same vasoconstrictor responses as mammals. The author has been unable to trace any convincing accounts of cold thermodilatation in birds, although there are a number of anecdotes which suggest that it exists. Ducks and geese held in warm conditions in autumn (therefore warm-acclimated) often freeze to the snow or ice if they are released outside during the winter. This indicates (a) that peripheral vasoconstriction is taking place, causing the feet to cool, and (b) that cold acclimated animals (which do not freeze to the ground) must allow enough blood to reach the feet to prevent freezing — this could be accomplished by reducing vasoconstrictor tone, but might be achieved by a regular vasoconstriction : cold vasodilatation cycle.

4.5 THERMOGENESIS

So far the emphasis in this chapter has been on the means by which endotherms minimize heat loss, but of equal importance is the means by which they generate heat. The basic features of heat production have been presented in Chapter 1, and the importance of brown adipose tissue to arousal in hibernators is considered in Chapter 5 in some detail. Attention is given here to the importance of shivering and non-shivering thermogenesis to normothermic animals (i.e. those maintaining a normal, high core body temperature) living in cold conditions.

For many high latitude endotherms, particularly the larger ones, there are no special metabolic or physiological mechanisms to generate heat; their anatomical and physiological insulation is so good (often reaching the equivalent of 5–8 'clo' units (Chapter 8)) that they have no trouble in balancing heat production and loss without resorting to enhanced shivering and non-shivering thermogenesis, except under the most severe weather conditions. The situation is different for birds and for the smaller mammals.

Birds generally appear not to have brown fat (the first observation of brown fat in birds was as recent as 1983 — by Oliphant in grouse and chickadees — and is so far based on histological rather than biochemical studies), and a number of investigations have indicated that the only way in which a bird can increase heat production at rest is by shivering (e.g. Palokangas and Hissa, 1971). Flying or swimming birds generate heat by action of the propulsive muscles. In the case of birds incubating eggs, heat loss is a particular problem because the eggs have to be kept warm by heat flowing through the brood patches (areas of well vascularized skin without feathers) into the eggs. Electromyographic investigations have shown that muscle activity (usually in the form of 'thermally-induced tremor', not readily observed visually) begins in birds as soon as the ambient temperature falls below t_c, the lower border of the zone of neutrality. Aulie and Tøien (1988) have recently demonstrated (in bantams,

Gallus domesticus) that aerobic muscles are the first to be involved in shivering (at ambient temperatures below 32°C), and such shivering can presumably be fuelled by oxygen and sugars/fatty acids for long periods. When ambient temperatures fall below 20°C, the bantam recruits anaerobic muscle fibres to the shivering process; presumably these muscles have limited ability to sustain activity. There is some evidence of special adaptation in high latitude birds. Bantams are basically derived from warm climate pheasant ancestors and have a mixture of aerobic and anaerobic fibres in their pectoral muscles. The willow ptarmigan, and other temperate/subarctic ground dwelling birds tend to have a greater proportion of aerobic muscle fibres.

Eider ducklings (*Somateria mollissima*) have attracted particular study. They hatch in the arctic and are precocial, being capable of swimming, eating and diving in water at 0°C within 24 h of hatching! When resting in water at 0°C their metabolic rate is 4 times that sustained within the thermoneutral zone (Steen and Gabrielsen, 1986), and they maintain normal body temperatures (approx. 40°C) throughout the body, despite their small size (approx. 60 g). Fine temperature measurements have demonstrated that only the leg muscles and liver have higher temperatures than the core tissues, with the thigh muscles being 2 deg. C warmer than the liver. Grav *et al.* (1988) have demonstrated that the thigh muscles are responsible for most of the thermogenesis. They also showed that there is an enormous increase in thigh muscle cytochrome oxidase activity during the 24 h period before hatching (Table 4.2), whereas activity in the liver is high during embryonic development. The enhanced activity of respiratory chain enzymes is matched by increases in

Table 4.2 Cytochrome c oxidase and tissue oxidative capacity of liver and thigh muscles in the Eider duck

Developmental stage	Liver (3.6 % body weight)		Thigh muscle (9.8 % body weight)	
	Cytochrome oxidase activity ($ng\ atoms\ O\ min^{-1}\ mg^{-1}$)	Oxidative capacity ($ml\ O_2\ g^{-1}\ h^{-1}$)	Cytochrome oxidase activity ($ng\ atoms\ O\ min^{-1}\ mg^{-1}$)	Oxidative capacity ($ml\ O_2\ g^{-1}\ h^{-1}$)
Embryonic (−24 h)	32	43	14	18
Newly hatched	34	45	50	67
Hatchling (+24 h)	40	54	52	69
Adult	35	47	11	14

Simplified from Grav *et al.*, 1988.

mitochondrial numbers and complexity of the mitochondria themselves (there is evidence of greater folding and packing of cristae, consistent with the hypothesis of Taylor (1987) that oxidative capacity is a reflection of the surface area of inner mitochondrial membrane per unit weight of tissue). Numerous lipid droplets were observed in close association with the mitochrondria, indicating involvement of free fatty acids in the intense metabolism.

The thigh muscles make up 9.8% of the body weight, the liver 3.6%. From the data shown in Table 4.2, Grav *et al.* (1988) calculated that the leg muscles of a 60 g duckling could, assuming no limitation of oxygen or biochemical substrate, have an oxidative capacity of 460 ml O_2 h^{-1}, the liver 119 ml O_2 h^{-1}. They recorded a maximum of 360 ml O_2 h^{-1} in an eider duckling exposed to water at 0°C, so it is evident that these thermogenic organs could produce more than enough energy to maintain a stable body temperature, but that the liver could not provide enough heat on its own. Grav *et al.* speculate that the rapid development of the ultrastructure of thigh muscles of eiders is triggered by enhanced output of the thyroid hormones triiodothyronine and thyroxine (presumably under pineal/pituitary stimulation). They present no direct evidence in support of this hypothesis, but such thyroid activity is seen in other precocial birds immediately before hatching. From Table 4.2 it may be seen that the high level of thigh muscle tissue activity is not maintained when the eiders grow; presumably adult animals, with their much more favourable surface-area to volume ratio can maintain body temperature without it, and, in any case, have large, functional pectoral muscles (undeveloped in the ducklings) which can supply shivering thermogenesis. It is not yet clear whether birds in cold climates derive all of the extra energy needed to maintain body temperature at low ambient temperatures from shivering thermogenesis; researchers are still searching for direct evidence of non-shivering thermogenesis by futile cycling or the uncoupling of mitochrondrial energy output from ATP storage.

In small mammals there is considerable evidence to show that brown adipose tissue (BAT) is an important source of emergency metabolic heat, not only in hibernating animals during episodes of arousal, but also in the case of normo-thermic animals during spells of low environmental temperatures. In many rodents, including rats, there are seasonal changes in the amount of brown fat present (usually in the neck region), with accumulation of BAT in the autumn. Warm-acclimated rats respond to cold conditions solely by muscular thermo-genesis. This is of limited effectiveness and warm acclimated rats showed a fall in core temperature when exposed to 10°C (Griggio, 1982). In contrast, cold-acclimated rats employ nonshivering thermogenesis (NST) to maintain a normal body temperature; they only shiver if noradrenaline release by the sympathetic system (which stimulates thermogenin synthesis in BAT, thus producing heat output) is blocked. Jansky (1973) reported that the intensity of NST in rats was inversely related to the acclimation temperature (acclimation or de-acclimation being accomplished in 3–4 weeks). The lower the acclimation temperature, the greater the metabolic response to noradrenaline injection. These observations suggest that mammals have much phenotypic scope for tailoring the amount of brown fat deposited to environmental conditions — an important adaptive feature in species with large zoogeographical ranges. Heldmaier (1971) and Jansky (1973) clearly demonstrated that the size of NST in comparison with basal metabolic rate was much greater in small mammals than larger ones. NST induced by noradrenaline injection in small species of bats could reach more than three times BMR; in rats and white mice nearly twice BMR, but in rabbits only 62% of BMR. Non-shivering thermogenesis triggered by noradrenaline release in adult animals as large as dogs and wolves is virtually negligible, indicating that anatomical and physiological insulation, combined with normal metabolic heat output and occasional shivering thermogenesis is adequate to sustain body temperatures, even under adverse conditions.

Young mammals often have poor temperature control, even in the case of quite large species. NST resulting from the mobilization of BAT has been implicated in thermoregulation in newborn members of a wide range of species, from rats and lemmings, to humans and even the harp seal (*Pagophilus groenlandicus*). Harp seal pups are born on ice floes and immediately encounter a thermal gradient of as much as 70 deg. C (Grav and Blix, 1975). They weigh 10–14 kg, but have minimal blubber, wettable infantile fur and no behavioural mechanisms to help them maintain body temperature. The mother may leave the pup for periods of time when the weather is cold. Grav and Blix demonstrated that as much as 10% of the body weight consists of brown fat, surrounding the body and richly supplied with venous plexuses (which presumably allow the rapid delivery of metabolic heat to the core). They also showed that maintenance of body temperature was dependent upon NST, but not dependent on shivering thermogenesis.

5 Sleep, torpor and hibernation

Both ectotherms and endotherms encounter lower temperatures at night, especially on land, than they do during the day. If they live at temperate or high latitudes they will also face lower temperatures in the winter than in the summer. These temperature differences tend also to be associated with fluctuations in food availability. For many daytime foragers, particularly birds, food is unavailable at night. This is especially true of nectar-eaters (e.g. humming birds) because flowers close at night but most insectivores and carnivores also gain the bulk of their food during daylight hours (bats are a conspicuous exception by virtue of their ability to find insect prey by echolocation). On a seasonal basis, winter in temperate and arctic latitudes is associated with a lack of leaves, fruit and seeds, together with a dearth of insects (which are generally in diapause).

Some animals cope with the fluctuations in temperature and food supply by migration to more equable climates, but for many of the smaller animals this is impossible and they adopt a variety of closely related strategies for energy saving which can be grouped together under the headings of sleep, torpor and hibernation. More extreme forms of energy saving which involve body temperatures well below 0°C, are dealt with in Chapter 6.

5.1 SLEEP

Regular periods of sleep are characteristic of birds and mammals. They are essentially periods of unconsciousness which are distinguishable from other forms of unconsciousness by the rapidity of arousal in response to various forms of disturbance (noise, vibration, changes in light intensity). Other vertebrate groups (e.g. turtles, crocodiles) show similar behaviour but whether they are truly asleep is a matter of some controversy amongst neurophysiologists. In this book, the precise definition of sleep is relatively unimportant, what matters is that it is a state in which metabolic rates fall to lower levels than would be sustained if the animal was awake. A wide variety of birds and mammals which spend their days in active foraging retire to safe nests, roosts, caves and dens at night where they sleep, often gregariously. In humans, sleep in a warm environment is characterized by oxygen uptake levels some 10% below 'basal metabolic rate' (the latter being measured in awake but inactive people). Probably sleep provides similar savings for roosting birds and sleeping small mammals, particularly for the latter which often have nest temperatures within the thermoneutral zone at night. A further factor which contributes to energy saving during sleep in endotherms is the circadian rhythm of body temperature fluctuation (usually by no more than 1–2 deg. C) seen in many birds and mammals (Aschoff, 1982). Periods of low body temperature normally coincide with periods of sleep (though rhythmicity appears to be controlled by photoperiod and not ambient temperature — Haim *et al.* 1988), which in turn tend to coincide with low ambient temperatures. A lowered body temperature, even by a small amount, permits the endotherm concerned to reduce the amount of heat production necessary to sustain a temperature gradient between the body and the environment. Assuming a Q_{10} of about 2 for metabolic rate, a 1–2 deg. C of decline of body temperature will 'save' some 7–15% of energy metabolism. Bakko *et al.* (1988) have recently investigated the black-tailed prairie dog, *Cynomys ludovicianus*. This colonial sciurid rodent (effectively a burrowing squirrel) is capable of daily torpor or even true hibernation, but neither of these responses are demonstrated unless conditions are extremely severe. Instead, the prairie dog shows pronounced diurnal body temperature fluctuations (<2.8 deg. C in winter) without torpor. The species also has a generally lower body temperature (by about 3 deg. C) in January than it does in August. Bakko *et al.* demonstrated that energy savings of around 28% accrue from these fluctuations. They also point out that a lowered nocturnal body temperature reduces evaporative water loss.

5.2 TORPOR AND HIBERNATION IN ECTOTHERMS

Torpor is a state of reduced activity, body temperature and metabolism. Torpor usually involves immobility, but not necessarily unconsciousness. Largely for convenience, short term (usually diurnal) torpor in ectotherms is usually considered separately from hibernation (prolonged torpor during winter). Because ectotherms can only control their body temperature by behavioural means (basically by exploiting the possibilities of basking in warm parts of the environment — Chapter 3), they are unable to avoid cooling at night or during winter periods when the sun is not available. Most terrestrial/freshwater ectotherms living in temperate latitudes will automatically become less active and have a lower metabolic rate at night, and the reduction in energy metabolism will be dependent upon the amplitude of temperature change. If the temperature change is great enough, the animal will become immobile. This sort of torpor, seen in insects, amphibia and reptiles, cannot in itself be considered adaptive in the sense that the torpor of humming birds is adaptive (section 5.3, p. 115). However, most ectotherms retreat to some form of cover as temperatures fall, thus avoiding predation and possibly restricting exposure to extreme air temperatures, so there are adaptive aspects of behaviour associated with torpor.

Hibernation in ectotherms has been relatively little studied in physiological and biochemical terms except in insects and reptiles. Most insects spend the winter in diapause, that is a state of arrested development (usually as eggs or pupae, but sometimes as larvae or adults). In diapause metabolic rate is very low (below the level expected from a fall in temperature) and useage of energy reserves slow. A few insects (e.g. queens of social insects such as wasps) overwinter in underground hibernacula as adults, and may survive for several years by virtue of repeated hibernation. Entry to diapause in insects occurs in response to shortening photoperiod, not to low temperature. This has the advantage that adverse conditions of temperature and poor food supply are anticipated. Light acts directly upon the insect brain which in turn affects the neurosecretory system (in aphids light may affect the neurosecretory cells directly). There are many varieties of diapause response, but in the giant silkworm *Hyalophora cecropia*, shortening daylength causes secretion of prothoracicotrophic hormone by the brain to be inhibited. This in turn results in suppression of ecdysone release by the thoracic glands, so development ceases and the insect spends the winter as a pupa. Because all growth processes are suppressed, metabolic rate is extremely low, particularly as all physical activity ceases too. Silkworms in diapause use no more than a few percent of the energy required to support growth and activity. During the winter, prolonged low temperature causes the brain to synthesize (but not release) prothoracicotrophic hormone. When temperatures rise in the spring the hormone is released into the blood, ecdysone release is stimulated and development rapidly proceeds to the adult stage.

In colonial insects other than honey bees (which maintain a high body temperature during winter by clustering — Chapter 3), the whole colony may hibernate. In ants of the genus *Myrmica*, the first responses to the problems of winter survival are seen in late summer (Brian, 1977). Larvae show reduced growth rates and divert energy into lipid storage; they enter diapause. The diapause larvae are carried deep into the soil by workers. The queen, who stops egg laying, soon follows. Late pupae are left near the surface in warm conditions so that metamorphosis into workers takes place before cold weather intervenes. Early pupae are killed and eaten. Over a period of several weeks the workers move into deeper regions of the nest and themselves enter a torpid state. It seems that metabolic changes take place which reduce oxygen uptake to very low levels, since ants will even survive flooding during hibernation. The end of hibernation is induced by sustained temperatures above about 8°C. Although the larvae are soon ready to feed, the workers and queens need a period of several weeks in which to develop their nursing behaviour, restart glandular activity and, in the case of queens, ripen their ovaries. Feeding of larvae and laying of eggs restarts when temperatures rise above about 10°C.

Most amphibia and reptiles living at temperate and high latitudes spend the winter in a torpid state. This is commonly known as hibernation, but evidence for responses other than a simple slowing of metabolism due to low temperature is limited, although Bennet and Dawson (1976) present evidence of such metabolic inhibition in a variety of temperate zone species.

Several tortoises are known to hibernate, including the Mediterranean species such as *Testudo graeca* and *Testudo hermanii*, the North American box tortoise *Terrapene carolina* and the desert tortoise *Gopherus agazzizi*. All show burrowing behaviour in response to shortening days and usually burrow to a depth where they will not encounter frost. The great naturalist Gilbert White (writer of *The Natural History of Selborne*, published in 1789) monitored the weight of his tortoise Timothy for 30 years and noted the weight decreases associated with winter dormancy between November and mid April (in Hampshire — hardly a natural habitat for a Mediterranean species!). All of the hibernating tortoises appear to lay down fat reserves prior to hibernation and this seems crucial to survival. Hibernating tortoises are immobile, unresponsive and clearly have a low metabolic rate. Arousal during winter is well known for *Testudo graeca* during warm periods; the aroused animals will feed and may not re-enter torpor in some cases. Arousal at the end of winter appears to be in response to rising temperatures rather than lengthening photoperiod. Hibernation seems to be a vulnerable time for tortoises; mortality in Carolina box tortoises is quite high, probably due to frost or inadequate energy reserves (since this species is unusually well protected against predation by its articulated shell). The physiology of hibernating *Testudo hermanii* was investigated by Gilles-Baillen (1979). She found that the animals show a great increase in blood osmolarity during hibernation. The tortoises' had a blood concentration of 250–300 mOsm l^{-1} in summer, but during hibernation this rose to above 450 mOsm l^{-1}. Much of the increase is due to accumulation of urea in the blood, but the data do indicate dehydration as well. A box tortoise (*Terrapene ornata*) provides some evidence of metabolic inhibition during hibernation which presumably adds to the energy-saving resulting from a low core body temperature. Glass *et al.* (1979) demonstrated a high Q_{10} of 3.7 between hibernating (5°C) and active (15–25°C) body temperatures. They also found high arterial pCO_2 levels. This respiratory acidosis has been implicated by Malan (1979) in metabolic depression (further discussed in relation to hibernation in endotherms).

Many lizards and snakes hibernate during winter, since temperatures below about 10°C are usually too low for effective food capture and digestion. A contributory factor may be that sluggishness at low temperature may make them vulnerable to predation, so that it is safer to retreat to hibernacula. Schwaner (1989) has recently studied the most southerly snake species, the black tiger snake *Notechis ater niger*; he found that specimens living isolated from predators on the Franklin Islands, South Australia did not hibernate, whereas mainland tiger snakes did.

Snakes inhabiting high latitudes (e.g. the adder, *Vipera berus*) routinely face long cold winters and may spend more than half of their life in hibernation (Gregory, 1982). Snakes store lipid in abdominal fat bodies and glycogen in the liver before the winter and then retreat, often communally as in the adder (Viitanen, 1967), to hibernacula (burrows, dens beneath stones, crevices etc). Mortality during hibernation is often high (18–47% in the case of adders according to Vittanen), but there is dispute about the causes of death. Gregory (1982) believed that exposure to lethal temperatures was the main cause, but other workers have reported that mortality stems from weight losses due to exhaustion of lipid reserves. For example, Charland (1987) reported weight losses of 25% in first-winter rattlesnakes (*Crotalus viridis*) during hibernation. On the other hand, it does appear that snakes often emerge from hibernation in good condition, still with sizeable fat bodies. Constanzo (1988, 1989) has recently carried out studies which strongly indicate that winter survival in several hibernating snakes is determined by access to water. For garter snakes (*Thamnophis sirtalis*) he reported high levels of survival in animals which spent the winter submerged in water, although able to breathe air at will (at 5°C all submerged animals survived for more than 165 days, while 50% had died after 133 days in air at the same temperature). In contrast, animals held in air showed pronounced weight loss. He found that mortality in all cases occurred when muscle tissue water had fallen to 65% of wet weight, even though substantial amounts of lipid remained. Clearly death in hibernation was due to desiccation rather than depletion of energy reserves. Constanzo also found that submerged snakes used lipid reserves at a low rate (0.38 g in a snake weighing about 28 g over 165 days, i.e. 2.31 mg d^{-1}), and that less than half of the lipid stores had been used up at this time. Assuming an energy value of 38.9 joules mg lipid^{-1}, a respiratory quotient for lipid of 0.72, and a consequent energy value of 19.8 joules ml O_2^{-1}, this corresponds to an oxygen uptake rate of about 0.007 ml O_2 g^{-1} h^{-1} (calculations by the author of this volume). If the hibernating snakes are only using lipid substrates, this oxygen uptake rate is probably about 1–3% of the oxygen uptake rate of active snakes in the summer (likely to be in the range 0.1–0.3 ml O_2 g^{-1} h^{-1}). Constanzo's studies also demonstrated that survival time is substantially reduced if hibernation temperatures are high. Garter snakes hibernating at 12°C used up fat reserves at more than twice the rate of those hibernating at 5°C; clearly it pays garter snakes to seek out deep, cold hibernacula to eke out their energy reserves. Overall, it is clear that selection of hibernation site is extremely important to reptiles hibernating in temperate and subarctic areas. Animals which choose shallow and dry hibernacula are much more likely to succumb to frost, energy depletion (caused by warming during short periods of winter sun), and to desiccation than are animals which overwinter in deep burrows or dens which feature stable, cold and damp conditions.

5.3 TORPOR IN ENDOTHERMS

Torpor in mammals and birds is characterized by a regulated reduction in body temperature and energy metabolism. It may be separated from sleep by the much longer period taken for normal activity to be resumed after stimulation. 'Regulation' is emphasized because, unlike ectotherms, torpid mammals and birds still regulate their body temperature above ambient levels, but at a level substantially below that associated with activity. In endotherms torpor is usually a nocturnal phenomenon, and is most common in small mammals and birds. Rough distinctions are made between the body temperatures of endotherms which show circadian rhythms of core temperature with minima corresponding to sleep (body temperatures not falling below 30°C), endotherms which enter torpor on a diurnal basis — 'daily heterotherms' (which usually have body temperatures during bouts of torpor between 10 and 30°C) and hibernating endotherms (whose body temperatures when torpid are usually no more than 2–4 deg. C above ambient, whether the ambient temperature is low (e.g. 5°C) or high (e.g. 20°C)). Daily torpor has been studied mainly in rodents, insectivores, small marsupials and humming birds. Examples of body temperatures and metabolic rates at normal and torpid body temperature are given in Table 5.1. From a more extended set of data Gieser (1988a) has concluded that animals exhibiting torpor on a daily basis do not exhibit any special reduction in metabolic rate beyond that attributed to simple temperature effects on metabolism. The Q_{10} values for daily heterotherms are usually close to 2, though there is an effect of body size, with small animals (e.g. shrews, body weight approx. 2 g) having higher Q_{10} values than bigger animals (e.g. badgers, body weight approx. 9000 g). Daily heterothermy seems to be limited to animals of less than 10 kg body weight and is an important adaptation for survival of food shortages and adverse environmental conditions in small endotherms (Hudson, 1978; Lyman *et al.*, 1982). Small animals cannot afford to carry the lipid reserves which allow larger animals to cope with short term shortages and low temperatures.

Small flying endotherms are particularly vulnerable to energy shortages. Daily torpor is especially important to the lives of many humming bird species (French and Hodges, 1959; Carpenter, 1974; Krüger *et al.*, 1982; Prinzinger *et*

Table 5.1　Body temperatures and metabolic rates of mammals and birds showing daily torpor. Basal metabolic rates correspond to the rate of normothermic animals (i.e. animals with normal, non-torpid body temperature). The Q_{10} values given correspond to the difference between basal and torpid metabolic rates over the temperature range between normal and torpid body temperatures

Species	Normal body temperature (°C)	Torpid body temperature (°C)	Basal metabolic rate $ml0_2\,g^{-1}\,h^{-1}$	Torpid metabolic rate $ml0_2\,g^{-1}\,h^{-1}$	Q_{10}
Marsupials					
Dasyuroides byrnei	34.3	24.4	0.74	0.44	1.68
Sminthopsis murina	35.0	15.0	1.13	0.25	2.10
Bats					
Nyctimene albiventer	37.0	28.6	1.43	0.67	2.46
Insectivores					
Suncus etruscus	34.7	14.0	5.75	0.60	2.98
Carnivores					
Taxidea taxus (badger)	37.0	28.0	0.26	0.13	2.15
Rodents					
Mus musculus (mouse)	37.4	19.0	1.47	0.65	2.30
Birds					
Calypte costae (humming bird)	40.0	21.0	3.0	0.39	2.90

Simplified from review of Gieser, 1988a.

al., 1986). Humming birds are amongst the smallest of living endotherms (body weight approx. 5 g) and may use oxygen at a basal rate of 3–5 ml $O_2\,g^{-1}\,h^{-1}$, little different from the metabolic rate of the smallest of mammals, the shrews. During flying their use of oxygen will be even greater. A number of humming birds live at altitude in the temperate zones, so encounter temperatures at least as low as 15°C. However, even those species living in lowland tropical environments often have daytime core body temperatures substantially higher than ambient levels. Schuchmann and Prinzinger (1988) have recently studied the Green Hermit (*Phaethornis guy*), a humming bird of Ecuador. The species lives in a thermally stable environment (27±3°C), but this is still an average of about 13 deg. C below the core body temperature of 40°C. During darkness the animal cannot forage and becomes torpid for a period of about 8 hours (Figure 5.1), during which the oxygen uptake rate falls from 8–10 ml $O_2\,g^{-1}\,h^{-1}$ to about 2–3 ml $O_2\,g^{-1}\,h^{-1}$, a saving of some 60–80%. When torpid and unable to feed the animal lost 7.5% body weight (presumably through a combination of respiratory loss of water and depletion of lipid reserves); this suggests that a

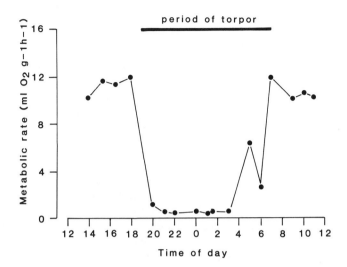

Figure 5.1 Changes in metabolic rate during nocturnal torpor in the green hermit (*Phaethornis guy*), an Ecuadorean humming bird. These data were collected from an animal held at 15°C; the body temperature during torpor was not recorded. Redrawn from Schuchmann and Prinzinger (1988).

non-torpid bird might easily lose more than a quarter of its body weight in a single night! Microchiropteran bats of temperate zones also employ daily torpor (Lyman, 1970) for energy saving, but in their case torpor occurs during the day, and is more profound than in most other daily heterotherms, since the body temperature in torpor is within 1 deg. C of ambient. In *Myotis lucifugus*, body temperature in daytime torpor may be as low as 7°C in summer (Lyman, 1970). Hill and Smith (1984) present an energy budget for the tiny Western pipistrelle (*Pipistrellus hesperus*) which weighs 3–5 g and demonstrate that these bats save 80–90% of energy metabolism by becoming torpid at 20°C.

Similar energy savings were reported by Collins *et al.* (1987) for a small marsupial, the honey possum (*Tarsipes rostratus*). This species is apparently an ideal subject for torpor studies. It is a nectar eater living in coastal heathland and shrublands in southwestern Australia. It is a small animal (5–15 g) and encounters both shortages of food (usually in autumn and late spring according to Collins *et al.*, 1987), and periods of fairly low temperature (2–6°C) throughout autumn, winter and spring. If nectar was available, the animals never became torpid, but if they were deprived of food the depth of the resultant torpor depended on the size of the animal and the coldness of the environment. Small size and low ambient temperatures resulted in deeper torpor.

All daily heterotherms lose and regain 'normal' (i.e. active) body temperature (normothermy) relatively quickly (1–3 hours) at the beginning and end of a bout of torpor. Relatively little attention has been paid to the mechanism of warming; as with hibernators it is probably due to a mixture of shivering and non-shivering thermogenesis, with brown fat playing an important part (except in birds which appear to rely only on shivering thermogenesis). Several studies have implied that control of cooling and warming is a rhythmic phenomenon related to photoperiod. This suggests an involvement of the pineal body and various hormonal systems, as is the case of hibernating endotherms in which it will be described in greater detail (section 5.4).

The occurrence, duration and depth of daily torpor are influenced by a number of factors. Generally speaking torpor is not employed by animals which are reproducing or lactating. Torpor is more frequent and more profound in winter than in summer (in non-hibernators) and both Collins *et al.* (1987) and Geiser (1988b) have demonstrated in marsupials that larger individuals of a given species tend to show shorter and shallower torpor than do smaller animals. Juveniles show longer and deeper torpor than adults; this is true of many placental mammals and birds too (Nagel, 1977; Prinzinger and Seidle, 1986). These observations reinforce the concept that daily torpor is an adaptation to cope with being small and having a high metabolic rate whilst living in an environment which features pronounced variations in temperature and food supply. Modern texts emphasize the positive nature of such adaptation which allows small endotherms to live in areas which would otherwise be inaccessible to them. This contrasts with older attitudes which categorized animals with stable body temperatures as more highly evolved than so-called 'lower' animals with variable core temperatures.

In considering the response of torpor it should be recognized that many small animals living at high latitude do not employ torpor at all (e.g. lemmings), or use it only under extreme environmental conditions (e.g. tree squirrels — Innes and Lavigne, 1979). On the other hand, some humming birds employ nocturnal torpor throughout their lives as a routine energy-saving strategy.

5.4 HIBERNATION IN ENDOTHERMS

Hibernation in mammals (and a few bird species) has attracted an enormous amount of study, from ecologists, physiologists, pharmacologists and biochemists. The corresponding literature is voluminous and beyond comprehensive review in a book of this nature. However, the following pages attempt a broad picture of current knowledge.

5.4.1 General characteristics of hibernation in endotherms

Hibernation ('winter sleep') is generally a feature of small mammals (<4 kg body mass), although a few birds (e.g. poorwills) hibernate, and a number of large carnivorous mammals (particularly bears living in subarctic and arctic regions) enter a dormant state in winter which is popularly regarded as hibernation, although some physiologists would prefer to avoid the term.

It is as well to point out that hibernation is not as common a response as is generally thought. Hibernation is rarely employed by small mammals living in arctic, subarctic and high altitude habitats. There are several reasons for this. First, low temperatures persist for so much of the year in these habitats (at least for 8–9 months) that hibernation is unlikely to be a viable option because small animals cannot store sufficient energy in the short summer to last for such a prolonged period, even though metabolic rates in hibernation may be low. Second, many species in these environments take advantage of the insulative qualities of deep snow and burrows to avoid exposure to very low temperatures; they also subsist on the vegetation and invertebrates in the subsnow microhabitat. Lemmings, snow voles, common and pygmy shrews (*Microsorex hoyi*) all fall into this category. Several of these species store vegetation in the short summer; Père David's voles, which live at an altitude of about 6000 m in the Himalayas are active during both day and night in summer, harvesting and drying the leaves and stems of vegetation for use in the winter, when they tunnel through snow to reach buried plants as well as their stored caches. Finally, quite a few of the small mammals found in these extreme environments (especially pigmy shrews and some of the smaller voles) have extraordinarily high weight-specific metabolic rates — the pigmy shrew has a heart rate of 1200 beats min^{-1} and needs to eat more than its body weight in a 24 hour period. Sustaining metabolism, growth and repair mechanisms is difficult enough for such animals in cold climates; storing energy for the future virtually impossible. Hibernation is therefore a feature of temperate zones rather than high latitudes.

It should also be realized that some animals which are popularly supposed to hibernate do not in fact do so. Grey and Red squirrels (*Sciurus carolinensis* and *Sciurus vulgaris*) are believed by most people in the USA and Europe to be hibernators, but both species are active throughout the winter and maintain high body temperatures. However, they spend more time in their nests, especially in cold weather, and rely on stored food rather than extensive foraging, so give an impression of disappearance and hibernation (the human tendency towards fewer country rambles in winter may also be a contributory factor to their 'disappearance'!).

Given these reservations, it is still the case that a wide range of small endotherms (including monotremes and marsupials, as well as placental mammals and birds) exhibit classic hibernation (Table 5.2). Hibernation of this type is a prolonged form of torpor in which core body temperatures fall to within a few degrees of ambient temperature (typically down to about 4–7°C in the coldest months). The hibernating animals maintain a low body temperature for periods of several months, but at intervals of about 1–2 weeks they

Table 5.2 Body temperatures and metabolic rates of mammals and birds showing hibernation. Basal metabolic rates correspond to the rate of normothermic animals (i.e. animals with normal, non-torpid body temperature). The Q_{10} values given correspond to the difference between basal and metabolic rates during hibernation over the temperature range between normal core temperature and body temperatures during torpid phase of hibernation

Species	Normal body temperature (°C)	Hibernation body temperature (°C)	Basal metabolic rate $ml0_2\,g^{-1}\,h^{-1}$	Hibernation metabolic rate $ml0_2\,g^{-1}\,h^{-1}$	Q_{10}
Monotremes					
Tachyglossus aculeatus (echidna)	32.2	5.7	0.15	0.066	1.36
Marsupials					
Cercartetus lepidus (pygmy possum)	33.7	10.1	1.49	0.055	4.00
Insectivores					
Erinaceus europaeus (hedgehog)	35.0	5.2	0.433	0.016	3.00
	35.0	16.0	0.357	0.011	6.20
Tenrec ecaudatus (tenrec)	33.0	16.5	0.27	0.025	4.20
Bats					
Myotis lucifugus	37.0	2.0	1.43	0.022	3.30
	35.0	5.0	1.53	0.060	2.94
	35.0	25.0	1.53	0.290	5.20
Rodents					
Glis glis (edible dormouse)	37.7	7.0	0.79	0.026	3.04
Spermophillus richardsonii (ground squirrel)	37.1	5.0	0.53	0.020	2.78
Birds					
Phalaenoptilus nuttallii (poorwill)	37.0	10.0	0.788	0.050	2.77
	37.0	20.0	0.788	0.086	3.68

Simplified from review of Gieser, 1988a.

briefly arouse, the body temperature rising within a few hours to normal levels, which are maintained for a short period (during which feeding and drinking may take place) before torpor is resumed. The interval between arousals tends to be short at the beginning of hibernation, lengthens and stabilizes for some months during the depth of winter, and then shortens again before hibernation ends.

The metabolic rate during torpor is extremely low. Gieser (1988a) tested the hypothesis that hibernators showed a physiological inhibition of metabolism, as well as a simple physical reduction in metabolic rate due to reduced body temperature. His results, derived from published information concerning a wide range of species (Table 5.2), tend to support this proposition, since higher Q_{10} values were recorded from hibernators than from daily heterotherms. He also demonstrated that the normal tendency for smaller animals to have a higher weight-specific metabolic rate than larger animals disappeared during torpor. This indicates that metabolic inhibition is more pronounced in smaller endotherms. He also showed that, within hibernation torpor, Q_{10} values decreased with decreasing body temperature (Table 5.2). This means that a hibernating endotherm does not increase its metabolic rate too much when ambient temperatures rise during the winter period, thus saving energy and avoiding inadvertent arousal. It also means that hibernating endotherms in torpor at relatively high ambient temperatures use less energy than daily heterotherms in torpor at the same body temperature (again indicating that torpor in hibernation is associated with metabolic inhibition).

5.4.2 Preparation for hibernation

Most hibernators prepare for winter by excessive eating (hyperphagia) during late summer and autumn. They take in more energy than is needed to sustain routine metabolism (and growth of lean body tissue in young animals). The excess energy is stored in two forms, as lipid in normal adipose stores ('white fat') and in the form of the lipid of hypertrophied brown adipose tissue (BAT — Chapter 1 provides an account of the biochemistry of brown fat tissue). These lipid reserves are the primary source of energy during hibernation, particularly for those animals, such as ground squirrels, which do not store food items for consumption during periods of arousal. Mature adult ground squirrels (i.e. animals showing negligible somatic growth), may increase in weight by 50–80% during August and September (Musacchia and Deavers, 1981). Hibernators which store winter food (e.g. hamsters) often do not put on weight in the autumn, but must forage for food items (particularly seeds, nuts) beyond their immediate needs to achieve the same ends. All hibernating small mammals possess quantities of BAT during the winter; it is crucial in the arousal process.

Reproductive and neurendocrine preparations for hibernation are somewhat confusing. There is a general tendency for regression of ovaries and testes, and at least in the Syrian ('Golden') Hamster (*Mesocricetus auratus*) it appears that atrophy of the gonads is important to hibernation. Smit-Vis and Smit (1963) reported that, if the pineal organ is removed, the gonads do not atrophy in the autumn (because the pineal antigonadotrophic hormone (melatonin) is not produced), and hibernation becomes impossible. However, the pineal (a combined light dosimeter and endocrine gland) is a controller of many seasonal processes, so the relationship between gonadal atrophy and subsequent hibernation may be associative rather than causal. In any case, other workers (e.g. Jansky *et al.*, 1981) have shown that golden hamsters do hibernate after pinealectomy, though to a lesser extent than in control animals! Castration promotes hibernation, but castrates which have also had the pineal removed show reduced ability to hibernate. There is also some evidence that the adrenal (suprarenal) glands are important in the induction of hibernation. Jansky *et al.* (1981) reported that adrenalectomy abolished the ability to hibernate in golden hamsters, while administration of deoycorticosterone and cortisol increased the frequency of hibernation in intact animals. There is a relationship between adrenal hormone levels and gonad development, so the adrenals may only be having an indirect influence, mediated through the gonads and secretion of sex hormones.

There is also a tendency for activity of the thyroid gland to be reduced or stopped prior to hibernation. However, Hudson (1981) points out that there is no consensus about the degree to which thyroid activity is reduced, or indeed whether activity is reduced before hibernation starts or falls in response to declining body temperature at the beginning of torpor. Thyroid hormones (thyroxine (T_4) and tri-iodothyronine (T_3)) are implicated in control of the intensity of energy metabolism (probably by regulating enzymes) and also in the stimulation of protein synthesis. Since both of these processes are suppressed during the torpid phases of hibernation it would seem to make sense that thyroid activity should be reduced during torpor. However, the available evidence is conflicting. Some hibernators switch off their thyroids (e.g. ground squirrels and marmots), others do not (e.g. hamsters). Even more confusingly, Hudson (1981) reported high plasma levels of T_3 and T_4 in ground squirrels during hibernation — though these levels could indicate an inability of tissues to take up the thyroid hormones during torpor. A further complicating factor is that low levels of thyroid activity have been associated with increases in the membrane fluidity of the tissues of hedgehogs, *Erinaceus europaeus* (Augee *et al.*, 1979). Since the tissues of hibernating mammals will fall to lower temperatures than are encountered by normothermic animals, it is an advantage for membrane fluidity to be increased before hibernation, but the mode of action of the thyroid in this respect is rather unclear.

5.4.3 Onset of torpor

At the beginning of hibernation endotherms cool over a period of roughly 1–3 days to the temperatures characteristic of sustained torpor. The animals cease movement and reduce sympathetic output so that the normal responses to falling temperature (peripheral vasoconstriction, shivering and non-shivering thermogenesis) are not deployed. The 'set points' of hypothalamic temperature regulation are reduced to levels a few degrees above ambient, while metabolic heat output is substantially reduced. Many species retreat to burrows or other hibernacula, and often seal the entrance(s). This tends to promote respiratory acidosis because of CO_2 build up — a factor in reducing tissue metabolic rate (Malan, 1979). Such acidosis appears to precede torpor. Respiratory acidosis also tends to depress the temperature at which shivering is elicited, so will help in the cooling process. Once the hibernating animal has entered torpor, the metabolic rate falls and relatively little CO_2 is produced, so the CO_2 within the hibernaculum will dissipate.

A question asked from the early days of study of hibernating animals has been 'is there a substance synthesized in the body which induces hibernation torpor?' Much early work centred around study of brown adipose deposits, which were once thought to have endocrine functions (see Swan, 1981 for review). It is now generally accepted that BAT is purely a thermogenic tissue, activated by neurendocrine control, but not itself elaborating hormones. Swan (1981) and Oeltgen and Spurrier (1981) present strong evidence for hibernation inducing hormone(s). Peptide fractions derived from the brains (particularly the hypothalamic regions) of hibernating ground squirrels are effective in inducing torpor in summer-active animals. They also appear to have appropriate physiological effects (e.g. reducing oxygen uptake and increasing coronary perfusion — section 5.4.4). Swan (1981) called the peptide fraction(s) 'antabolone' and described it/them as an antimetabolic neurohormone or group of hormones. Plasma fractions in the 5000 dalton range taken from hibernating woodchucks (*Marmota monax*) and 13-lined ground squirrels (*Citellus tridecemlineatus*) induce summer hibernation in the same species, between the two species and also caused torpor and core temperature drop in macaque monkeys (*Macaca* spp. — which are non-hibernators). Oeltgen and Spurrier (1981) also showed that the plasma-borne hormone was not present in the blood of arousing hibernators. This observation suggests that (a) the hibernation inducing hormone system is switched off during arousal; (b) that it must be switched on again for the animal to re-enter torpor. It seems likely that there is a relationship between photoperiod, pineal activity and secretion of a hibernation inducing hormone and such a relationship could help to explain how a period of hibernation ends as well as how it begins. Clearly this is an exciting area of pharmacological study, made complicated by the wide range of peptides found in brain tissue.

5.4.4 Physiology and biochemistry during torpor

During torpor, hibernating endotherms have low metabolic rates, low body temperatures and cannot drink. The first feature is obviously adaptive as it allows the hibernating animal to save energy at a time of the year when energy is difficult or impossible to collect. Low body temperatures and lack of access to water pose physiological and biochemical problems. If a non-hibernating endotherm has its body temperature lowered below about 20°C it will suffer heart failure due to fibrillation of cardiac muscle. Although peripheral body tissues of non-hibernating mammals are often resistant to prolonged exposure to low temperature, this is not the case for deep tissues — so why do hibernators not suffer heart failure and cold injury? Some small desert rodents (e.g. jerboas — *Jaculus jaculus*) do not need to drink water, but most normothermic endotherms, particularly those feeding on seeds and nuts, require frequent drinks — so how do hibernators avoid or cope with dehydration?

The energy savings accrued during the torpid phases of hibernation can be very great (Table 5.2). The lower the hibernation temperature and the smaller the animal, the more energy is saved. Echidnas (about 2800 g body weight) hibernating with a core temperature of 5.7°C save only 56% over their normothermic basal levels, whereas hedgehogs (700 g body weight) save 96% when their body temperature is lowered to 5.2°C. Bats can achieve phenomenal savings — *Myotis lucifugus* (5–6 g body weight) hibernating with a body temperature of 2°C save 98.5% over normothermic basal rates. Because hibernating animals have a very low respiratory rate they lose little respiratory water. Reliance on lipid reserves reduces the need for urinary water loss and provides metabolic water, so depletion of body water is not normally a problem during torpor.

Cardiovascular responses during the torpid phases of hibernation have attracted attention from two points of view. First, how does the heart function at low temperature, and second, what are the consequences of reduced heart beat? The 13-lined ground squirrel (*Spermophilus tridecemlineatus*) has provided much useful information about cardiac function. Normothermic ground squirrels ($T_b = 37$°C) have a heart beat of about 240 beats min^{-1}. During hibernation at a T_b of 5–7° the heart rate drops to 3–4 beats min^{-1}. When the heart rate is reduced, the cardiac output is decreased, but aortic pressure is maintained and there is a progressive direction of a greater proportion of blood flow to the coronary arteries — so that the heart muscle is perfused effectively, allowing the heart to avoid fibrillation (Burlington and Darvish, 1988). It has been noted that the ground squirrel cardiac muscle has an unusual concentration of lipid droplets associated with the mitochondria (Burlington *et al.*, 1972) and it appears that these droplets provide free fatty acid fuel for the myocardium. Burlington and Shug (1981) provided further support for this hypothesis when they demonstrated that ground squirrel myocardium contains carnitine (essential for free fatty acid translocation to the mitochondria). Similar provision of fuel and oxygen to heart muscle is not seen in non-hibernators such as laboratory rats.

Aortic blood pressure is maintained in hibernation at a reasonably high level and perfusion of heart, brain, liver and brown adipose tissue continues efficiently throughout torpor. To accomplish this, much of the peripheral circulation of blood is greatly reduced by peripheral vasoconstriction (Lyman, 1982), which indicates a degree of sympathetic control (by adrenaline and noradrenaline). Since sympathetic output must be depressed during the cooling period at the onset of torpor, it must be stimulated again as the body temperature falls to the levels maintained during torpor. The reduced blood flow to the tissues, a low respiratory rate (most small hibernators breathe at little more than 1–2 breaths min^{-1}) and confinement to hibernacula interact to produce considerable respiratory acidosis because of the build up of CO_2 (see Malan, 1977, 1979, 1982, 1986; Malan *et al.*, 1973). Respiratory acidosis causes the blood pH to be lowered. In the case of European hamsters (*Cricetus cricetus*) the normothermic plasma pH and pCO_2 values are about 7.4 and 60 torr respectively. During deep hibernation the pH declines to 7.0 and the pCO_2 rises to 160 torr (Malan *et al.*, 1988). Changes of pH have profound effects on enzyme activities. For example: a decrease of pH by 0.1 of a unit causes a tenfold decrease in the activity in phosphofructokinase (Trivedi and Danforth, 1966). Malan (1979) reported that pH was depressed in most tissues (except the heart and liver) during hibernation. The pH depression causes considerable metabolic inhibition (thus explaining the 'extra' energy saving of hibernation torpor by comparison with daily heterothermy). The pH of liver and heart is maintained at high levels (at the expense of energy-consuming proton pumping) to permit constancy of enzyme functioning in these crucial organs.

Torpid hibernating endotherms maintain low tissue temperatures for long periods. In this they only differ from other cold climate endotherms in that such temperatures extend to the body core of heart, brain and liver instead of being limited to the periphery. This means that membrane characteristics in terms of fluidity and exchange functions (mainly involving membrane-bound proteins) must be maintained over a wide range of temperatures. There must therefore be a degree of seasonal biochemical acclimation of the type described for ectotherms (Chapter 1), and enzyme function has to be strongly regulated to maintain function over a wide range of body temperatures (Malan, 1979).

5.4.5 Arousal from torpor

One of the most fascinating features of hibernation is the periodic arousal from torpor. Within a few hours (or even minutes in the case of the smallest hibernators such as bats) the hibernating animal is transformed from a motionless state in which the body temperature is close to ambient (perhaps as low as 2–5°C) and external signs of life are almost absent, into a fully active creature with a body temperature of 35–40°C. Much evidence is now available to tell us how the warming up and restoration of function are produced, but the reasons for periodic arousal are still rather obscure (Gieser and Kenagy, 1988). Two main and one minor hypotheses have been offered. The first is based upon the concept that hibernation patterns are exaggerated versions of the endogenous rhythmic sleep : activity cycle of endotherms. Torpor represents sleep, arousal represents wakefulness (e.g. Folk, 1957; Lyman *et al.*, 1982). The second hypothesis is that arousal is a response to depletion of energy stores and build-up of metabolic wastes (e.g. Mrosovsky, 1971). In the case of bats it has also been suggested that bats arouse to replenish body water or to move to more favourable locations (Hill and Smith, 1984). None of the hypotheses is entirely satisfactory, principally because arousals are expensive procedures in energetic terms, and it is clear that most energy consumed during the whole hibernation period is used up during the episodes of arousal. It is particularly difficult to accept the first proposition; that arousals are simply a manifestation of an endogenous rhythm. Surely the selection pressures to eliminate arousal for conservation of energy would have been extremely strong? The second hypothesis has much more to commend it, and certainly would be applicable to those species which have access to stored food (a majority of the small herbivores) or which are capable of foraging when aroused (e.g. bats, hedge-hogs). More difficult to deal with are the cases of hibernators which rely on fat stores laid down in the autumn to fuel the whole of hibernation (mainly the larger herbivorous forms such as marmots). In these cases it would seem to be maladaptive for animals to respond to energy depletion by arousal. Dealing with accumulated waste products is conceivably the main stimulus for arousal in these animals, but the recent discovery by Zhegunov *et al.* (1988) that intense and widespread protein synthesis is associated with arousal in gophers, suggests that repair and maintenance of functional and structural proteins may be an important function of arousal. A further reason for arousal has often been suggested, particularly in the older literature. This is, that hibernators are stimulated to arousal by falling environmental temperatures, and thereby avoid succumbing to death by freezing. There is some laboratory experimental evidence to support this, but it is rather equivocal, and may in any case be a laboratory artifact, since most hibernators choose their hibernacula with care, thereby avoiding extremely low temperatures. Energetically the concept is dubious, since the aroused animals would be confronted with a great thermal gradient between environment and body core at a time when foraging would be virtually impossible. However, arousal in response to adverse environmental conditions may be of value to bats, which sometimes hibernate at the mouths of caves, so perhaps gain benefit by arousing and moving to more favourable sites deeper in the cave systems.

Little is known of the triggers which initiate arousal, but initiation is presumably produced by the neurendocrine system, and may well involve a release of an antimetabolic inhibition produced by 'hibernation inducing hormone(s)' (section 5.4.4).

Arousal has several components, the first being the reversal of the general acidosis associated with hibernation torpor; metabolism cannot be stimulated until the metabolic inhibition caused by low tissue and body fluid pH is overcome. Malan *et al.* (1973) found that the first sign of impending arousal is the development of an increased breathing rate (hyperventilation). Hyperventilation involves ventilation of the lungs at a rate greater than is needed to sustain oxygen consumption, thus 'washing out' CO_2 from the body and raising pH levels. In the case of hibernators in torpor, the oxygen uptake is very low, so a modest increase in breathing rate causes a massive relative hyperventilation, without involving great expenditure of energy in fueling the breathing movements of ribs and diaphragm.

Next the animal begins to warm up (Figure 5.2), usually in an exponential fashion. In mammalian hibernators the initial stages of the warming process are produced by non-shivering thermogenesis (NST), much of it apparently sited in brown fat adipose (see Jansky, 1973; Musacchia and Jansky, 1981 for review). NST provides around 40% of heat during the arousal of hamsters (Hayward and Lyman, 1967) and BAT provides much of this thermal energy.

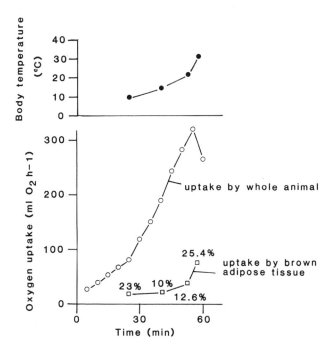

Figure 5.2 Oxygen consumption of bats during arousal from hibernation torpor. The percentage values for the brown adipose curve represent the maximum contribution that the brown adipose tissue can make to total metabolic demand during arousal. Clearly BAT is important to arousal, but other sites of muscular and non-muscular thermogenesis must be involved. Redrawn from Jansky (1973).

There is considerable evidence to show that BAT is stimulated to activity by noradrenaline and adrenaline secreted from the adrenal medulla and other chromaffin tissue under sympathetic control deriving from cells in the hypothalamus. The thyroid hormones also stimulate heat production by BAT and their release is also under hypothalamic control (via the anterior lobe of the pituitary gland).

Once some warming has taken place the animal begins to warm up more quickly, mainly by the action of contracting locomotor muscle (shivering thermogenesis). Blood insulin and blood glucose levels rise rapidly, indicating mobilization of glycogen to support the increasing muscular activity. Insulin secretion in the arousing hedgehog does not start until the body temperature has risen to about 25°C (Hoo-Paris *et al.*, 1978), so the initial contribution of NST arousal is clearly crucial. However, there are other sources of heat available as body temperatures rise. Burlington and Darvish (1988) have recently investigated cardiac function in the 13-lined ground squirrel (section 5.4.4). They considered cardiac efficiency (energy output/energy input) at various heart temperatures. Interestingly, the efficiency is greatest (15–16%) at low body temperatures (7–17°C), but declines to 10–11% when body temperatures rise to 27–37°C. In contrast, the non-hibernating rat shows high efficiencies (17–20%) at high body temperatures (27–37°C). Burlington and Darvish believe that the decreasing efficiency of cardiac metabolism in the ground squirrel as body temperatures rise above the levels characteristic of hibernation torpor means that the heart itself contributes to non-shivering thermogenesis during arousal; in effect it is a heat producing organ which will produce more heat as the body temperature rises, the cardiac efficiency falls, yet the heart rate increases greatly. It is not known why less energy is devoted to cardiac work and more released as heat — there is no evidence that the hearts of hibernators have an uncoupling of mitochondrial respiration from ATP synthesis in the manner of brown adipose tissue. However, Zhegunov (1987), working upon another ground squirrel species (the long-tailed gopher, *Citellus undulatus*), reported considerable changes in the ultrastructure of cardiac muscle cells during the process of arousal. He found marked increases in the number of elements of the rough endoplasmic reticulum, Golgi complex, vesicles and ribosomes around the nucleus. These areas contain very high concentrations of mitochondria and lipid droplets, so this structural investigation reinforces the concept of the heart pouring out heat (as well as blood) during the arousal process. It is even conceivable that the high heart rates seen towards the end of arousal (higher than occur during normal activity) may be partly thermogenic rather than being required for purely circulatory purposes.

Zhegunov *et al.* (1988) have recently found that there is a profound increase in protein synthesis during arousal in most of the tissues of long-tailed gophers, even though these tissues have greatly varying functions. Protein synthesis, concentrated in the membrane fractions of microsomes and mitochondria, is about 3 orders of magnitude greater than during hibernation torpor (in which protein synthesis is greatly depressed, to around 0.1–0.5% of the level recorded in active gophers). Zhegunov *et al.* interpret this acceleration of synthesis on arousal as a response to replace functional and structural proteins lost or altered during hibernation torpor. However, the intensity of synthesis (perhaps 1–2 orders of magnitude greater than in normothermic gophers) suggests that it may have much to do with supporting the generally high level of metabolism sustained during arousal (roughly 40 times the metabolic rate seen at the same body temperatures during the cooling phase at the onset of torpor) by producing changes in the ultrastructure in a manner which promotes energy translocation.

Very few bird species enter seasonal prolonged torpor, and all appear to belong to the family Caprimulgidae (goatsuckers, poorwills and nighthawks) (Dawson and Hudson, 1970). It would appear that arousing poorwills (*Phalaenoptilus nuttallii*) have to rely on passive environmental heating to bring the body temperature up from 5 to 15°C; thereafter the animals arouse by vigorous shivering — there is no evidence of NST.

So far only endogenous sources of heat have been considered. Many hibernating mammals are social and hibernate in clusters (Chapter 3). Close physical contact is counterproductive at the onset of torpor, but Arnold (1988) has recently demonstrated that alpine marmots (*Marmota marmota*) gain thermal benefits by synchronized arousal. The heat of normothermic, aroused animals helps passively to rewarm their torpid nestmates. Awakened marmots will deliberately huddle with torpid animals, and may also cover them with vegetation to improve insulation.

5.4.6 Termination of hibernation

As spring comes, the periods of hibernation torpor become shorter and the periods of activity after arousal longer (Twente and Twente, 1967). This may reflect lengthening photoperiod, rising temperatures or increased food availability, but it is probable that all three factors interact until the production of hibernation inducing factors/hormones ceases until the next autumn.

5.5 SUPERCOOLED HIBERNATION IN ENDOTHERMS

The majority of small mammalian hibernators maintain core body temperatures of at least 2–4°C when torpid. Usually they live in borrows or nests so that the ambient temperature is little below body temperature. Many years ago it was demonstrated that hibernating bats may be exposed to temperatures at least as low as −17°C (Mislin and Vischer, 1942), because bats often overwinter in relatively poor shelter (Davis, 1970). Body temperatures in several chiropteran species (e.g. *Plecotus auratus, Nyctalus noctula, Myotis daubentoni, Myotis lucifugus, Myotis sodalis* and *Lasiurus borealis*) are allowed to fall below the melting point of blood plasma (roughly −0.6°C), so that the animals are supercooled (Eisentraut, 1934; Kalabuchov, 1935; Davis and Reite, 1967). It appears that supercooling to −5°C occurs by allowing passive cooling. If ambient temperatures are below −5°C the animals respond by increasing the metabolic rate (e.g. *Lasiurus borealis, Nyctalus noctula*). The phenomenon of supercooling in bats was studied in most detail by Davis and Reite, who found that if *Lasiurus borealis, Myotis sodalis* and *Nyctalus noctula* were kept at −5°C they first thermoregulated, maintaining a heart rate of about 200 beats min^{-1}. When they ceased thermoregulating the rectal temperature rapidly fell to ambient (i.e. −5°C) and the heart rate dropped to about 9 beats min^{-1}. This heart rate compares with a heart rate in bats hibernating at +5°C of 24–34 beats min^{-1} and roughly indicates that supercooled hibernation involves an energy saving of 60–80% over 'normal' hibernation. The supercooled bats appear not to breathe and heart activity is faint as well as slow, so the energy saving may be much greater. Davis and Reite demonstrated that, in *Myotis lucifugus*, there is a three part response of bats to low temperatures in hibernation. If a bat is hibernating at an ambient temperature of 10°C it will be torpid, with a heart rate of about 50 beats min^{-1}. As temperatures fall to +5°C the metabolic rate drops, the heart rate falling to 25 beats min^{-1}. At temperatures below 5°C the bats respond by complete arousal (unless the rate of cooling is slow, in which case they may remain torpid), the heart rate rising to 200 beats min^{-1} at a temperature of −5°C (Figure 5.3). After a few hours of arousal the animal ceases thermoregulation and supercools to ambient temperatures (heart rate 9 beats min^{-1}). There is good evidence that heart rate and metabolic rate are closely correlated in bats, so the decrease in heart rate suggests that *Myotis lucifugus* 'saves' about 95% of energy expenditure by switching from thermoregulation to supercooled hibernation. The supercooled state can last at least for several hours and warming the animal to +5°C restores the heart rate to normal hibernating levels.

That the bats are supercooled has been proved by inducing freezing by pricking the animals with a needle; the core body temperature immediately rises to a plateau at −0.5 to −1.0°C as the animal is warmed by the heat of crystallization. No evidence appears to have been presented to suggest that bats possess any antifreeze or cryoprotective agents. Davis and Reite suggested that supercooled hibernation in bats was only of significance during short term exposure to subzero conditions, perhaps when wind direction changed at the mouth of the cave used for hibernation.

Barnes (1989) has recently demonstrated that the arctic ground squirrel (*Spermophilus parryii*) is also capable of supercooling, but in this case for long periods. Most arctic mammals do not hibernate because the winters are too long and severe. The arctic ground squirrel is an exception and may hibernate

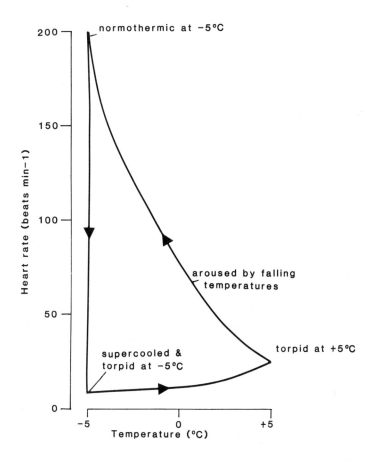

Figure 5.3 Heart rate of the bat *Myotis lucifugus*. This bat hibernates regularly with body temperatures down to 5°C. If the environment of the bat cools below this level, the bat arouses and shows greatly elevated heart rate (about 200 beats min^{-1}) if temperatures fall to −5°C. The bats remain normothermic at this temperature for some time, but then become torpid; body temperature drops to near ambient levels and the heart rate falls below 10 beats min^{-1}. From data of Davis and Reite (1967).

for 8–10 months each year in hibernacula in soil as cold as −18°C. Barnes monitored deep body temperatures in arctic ground squirrels by implanting temperature-sensitive radiotransmitters (Minimitters) and found that they had deep body temperatures as low as −2.9°C, yet did not freeze. In his experiments the hibernaculum temperatures were never below −7°C, so body temperatures may be lower in some natural burrows where much lower temperatures were recorded. As with the bats described above, there was no evidence of plasma antifreezes, but in this case experimental evidence actually excluded the presence of such molecules since the freezing and melting points of blood plasma were virtually identical at -0.59 and $-0.56°C$ respectively (i.e. there was no thermal hysteresis). Body temperatures below the freezing point of the blood were often maintained for 2–3 weeks between episodes of arousal, and sequences of supercooled torpor interspersed with brief arousals lasted for months. Clearly existence in this metastable state is a normal feature of the species' biology.

Barnes also showed that there are thermal gradients within the body of a supercooled arctic ground squirrel. Although the posterior region of the body is supercooled, with lowest temperatures being recorded in the colon and feet, the anterior portion of the body is some 1–2 deg. C warmer. Keeping the brain, heart and brown adipose tissue rather warmer than the rest of the body is presumably of physiological significance, perhaps because it allows blood circulation to continue. However, the heterogenous distribution of body temperatures makes evaluation of the energetic value of supercooled hibernation difficult. Barnes suggested a ten-fold advantage over maintenance of normal hibernation temperatures (i.e. $>0°C$), but admitted that this might be an overestimate.

In both bats and ground squirrels supercooled hibernation seems to be an adaptation to prevent lethal energy consumption under extreme conditions of temperature which are avoided by most other hibernators. In neither case is it clear why freezing does not occur. Other hibernating mammals (e.g. rats, hamsters) have been induced to supercool under artificial conditions (reaching body temperatures as low as $-5°C$), but in these cases supercooling has ended with spontaneous freezing after less than an hour (causing damage and usually death). Photographs of hibernating bats show that their fur may be covered in frost, so it seems improbable that they, or *Spermophilus parryii*, are not supercooled in the presence of ice crystals. There is increasing evidence that penetration of some sorts of structures (capelin egg chorions, gammarid integuments, ice fish corneas and urethras) by ice nuclei is difficult, but the basis of maintenance of the precarious state of supercooling clearly merits further study.

5.6 HIBERNATION IN BEARS

A number of bears of the genus *Ursus* (e.g. brown, grizzly, Himalayan) living at high latitudes and/or high altitude retreat to dens in the winter. Some bears shelter in natural caves, but in many cases dens are dug out of hills, or excavated beneath the roots of big trees. Dens are large, being 3–4 metres deep and may contain spruce or fur bedding. Female polar bears (*Thalarctos maritimus*) dig large, chambered snow dens, but only in winters when they are to produce cubs. The depth and size of dens plus the great bulk of the animals themselves means that the animals will spend the winter in a relatively warm, humid microclimate, quite different from the cool conditions encountered by hibernating small mammals. There is even some evidence that grizzlies in Colorado deliberately choose slopes where thick snow is likely to improve the den insulation. Brown bears and various Himalayan species may spend 3–5 months in their dens, giving birth to cubs whilst denned up; polar bear females may be in their snow holes for as much as six months. There is no real evidence of photoperiod inducing denning up — it appears that the bears remain active in the autumn until food becomes scarce. All of these bears show hyperphagia (excess eating) in the summer and early autumn, so lay down considerable fat stores before entering a period of dormancy which has been variously termed 'heavy sleep' and 'carnivorean lethargy' (Lyman and Chatfield, 1955; Hock, 1957), although the term hibernation is still widely used. Brody and Pelton (1988) have recently investigated seasonal changes in the digestion of black bears (*Ursus americanus*). They found that bears ate nearly twice as much in November as in August. They also found a shift in gut assimilation of material in November by comparison with the summer; bears preferentially assimilated fat and carbohydrate in autumn at the expense of protein.

It has long been known that torpor is very shallow in hibernating bears. Hock (1951) reported that decreases in body temperature during hibernation were slight; core temperatures fell from 37–39°C in summer to 31–35°C in winter. Nelson *et al.* (1973) found that bears spent much of their time apparently asleep in their dens, curled up in a nest of vegetation. However, the bears were alert enough to attempt occasional attacks on the experimenters! These latter workers carried out a range of physiological studies which have provided most of our knowledge of bear hibernation. They established that brown bears lost about 25% of their body weight during 100 days of hibernation, but were in good physiological condition at the end of this period (wild female bears are capable of suckling their cubs even before they have left the den). Nelson *et al.* also established that hibernating bears neither defaecated nor urinated, thus minimizing water loss. Maintenance of a curled-up posture within a humid den would also reduce respiratory water losses to a low level. Whether bears have a suppressed ventilatory response to heightened inspiratory pCO_2 seems not to be known, but is highly likely. Estimates of the energy saving achieved during hibernation are difficult. All of the high latitude bears are perfectly capable of surviving winter temperatures indefinitely by virtue of their bulk, fur and cardiovascular physiology; it is the lack of food supply (or need to be on land for the birth of cubs in the case of the female polar bear) that dictates dormancy. It is therefore difficult to decide what the metabolic rate during hibernation should be compared with. In most well insulated dens it is likely that the bear is living within its thermoneutral zone and perhaps saving 30–40% over summer basal metabolic rate (assuming a 4–5 deg. C fall in body temperature and a Q_{10} of 2). However, the

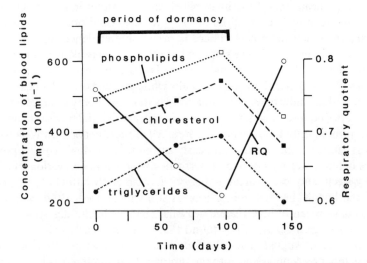

Figure 5.4 Changes in lipid metabolism in bears (*Ursus americanus*) during winter dormancy ('hibernation'). There is a steady rise in circulating lipids (and a concomitant decline in respiratory quotient) during dormancy as the bears avoid breaking down protein. Redrawn from Nelson *et al.* (1973).

saving over metabolic rate during active autumn foraging must be very much greater.

The observation that bears did not urinate or defaecate for several months raises interesting physiological questions which Nelson and his co-workers addressed. First, they established that hibernating bears progressively switched to a lipid based metabolism (Figure 5.4) and that the respiratory quotient consequently dropped throughout dormancy, while circulating plasma levels of triglycerides, phospholipids and cholesterol rose. Evidently the bears maintain protein catabolism at a low level, so need to produce little urine. Table 5.3 shows that there is a dramatic drop in urine volume and urine nitrogen during hibernation. However, small amounts of urine are still produced continually, yet the bears do not urinate and retain relatively small amounts of urine within the bladder. It appears that the urinary bladder is a site of water and nitrogen reabsorption, the absorbed nitrogen being disposed of metabolically (since there is little alteration in blood urea, uric acid or ammonia during hibernation, only creatinine showing significant increases). Urine production and urine reabsorption are balanced so that urine volume in the bladder does not rise sufficiently to stimulate bladder stretch receptors and thus trigger urination. Reduction in urea synthesis in response to food deprivation appears to be the most important physiological response

Table 5.3   Urine volume and urine nitrogen before and during hibernation in brown bears, *Ursus arctos*

	Before hibernation	*During hibernation*
Urine volume (ml d^{-1})	1872	107
Total N$_2$ (g d^{-1})	15.5	3.5
Uric acid (mg d^{-1})	543	133
NH$_3$ (mEq d^{-1})	12.2	6.1
Creatine (g d^{-1})	2.78	2.46

These data show that urine is much more concentrated during hibernation and that there are marked reductions in ammonia and uric acid excretion, though creatine output remains unchanged
After Nelson *et al.*, 1973.

in the maintenance of winter dormancy in bears, but is simply a more extreme version of responses seen in other mammals (including humans).

Although bears are thin and hungry at the end of hibernation, the evidence of Nelson *et al.* indicates that they are otherwise in good condition, with indicators such as plasma ions, thyroid hormone levels and blood glucose concentrations all being close to levels seen before hibernation. They are not dehydrated, which is particularly important in the case of nursing females; obviously recycling of body water is mainly responsible for this, but the microclimate within the den (plus metabolic water?) presumably plays a part.

PART THREE

Life at Temperatures Below 0°C

6 Subzero survival in terrestrial animals

6.1 SURVIVAL OF INSECTS AND OTHER TERRESTRIAL ARTHROPODS

Although maximum diversity amongst terrestrial arthropods (insects, spiders, scorpions mites, ticks) occurs in the tropics, and relatively low numbers of species occur in high altitude, high arctic and antarctic regions, the few cold-tolerant species are often important ecologically in cold areas. Anyone who has encountered the high biomass of biting flies in marshy areas of Canada, Iceland, Norway and Finland can certainly vouch for their prominence! Because such arthropods (particularly insects and mites) are characteristic of inland ecosystems (e.g. tundra, muskeg), far from the moderating influence of the sea, they may encounter very low temperatures (-30 to $-60°C$) in winter. Because the ground and vegetation freezes it is often impossible for these animals to avoid exposure to the full rigours of ambient air temperatures. Some arthropods maintain activity at subzero temperatures; Dalenius (1965) reported that the mite *Maudheimia wilsoni*, which tolerates temperatures as low as $-30°C$, was capable of locomotion and reproduction at temperatures below $0°C$. However, the majority of temperate and high latitude arthropods spend the cold winter in diapause — a state of arrested development or dormancy.

Diapause has been studied best in insects in which the life cycle is adapted so that the growing phase (larva, nymph) coincides with an adequate food supply. To ensure this, diapause can affect virtually any stage of the life history (egg, larva, pupa, adult/imago), depending on the species concerned. To minimize energy wastage, diapause is characterized by extremely low levels of metabolic rate.

Control of the onset and termination of diapause has been much studied, particularly in large insects such as locusts (*Schistocerca*, *Nomadacris*) and it seems that the main environmental factor involved is that of daylength (photo-period). Typically, at high latitudes, shortening photoperiods induce diapause; increasing daylengths in spring bring it to an end. Photoperiod seems to be the dominant controlling factor because it allows the arthropod concerned to anticipate adverse conditions (i.e. diapause can start before temperatures fall). However, diapause may also be affected by temperature and food supply. Diapause induction and termination are mediated through the central nervous system; either light directly affects the brain, which stimulates neurosecretory cells to elaborate a diapause hormone, or light may even have a direct effect on neurosecretory cells (as in aphids). It is also known that diapause induction may be started in one phase of the life cycle (e.g. the adult), but expressed in another (e.g. the egg).

Entering diapause during winter may solve problems of energy shortage in cold, winter weather, but it does not explain how insects and other arthropods tolerate temperatures well below the expected freezing point of most biological fluids. Cold-hardiness in terrestrial arthropods has attracted much study by physiologists and biochemists, but before considering the mechanisms involved, two features of arthropod anatomy need to be emphasized, since they have an important bearing on the ability of insects and mites to tolerate freezing temperatures; they are preadaptations to survival.

First, terrestrial arthropods are characterized by the possession of a non-living chitinous cuticle, usually coated with a waxy outer layer (epicuticle) secreted by glands in the epidermis. The waxy layer minimizes desiccation and also tends to make the cuticle hydrophobic. The existence of a non-living, hydrophobic barrier between the tissues and the environment helps to make the nucleation of the interior of eggs, larvae, pupae or adults by ice crystals rather difficult.

Second, and more importantly, terrestrial arthropods, particularly insects, have evolved a very different respiratory system from other terrestrial animals. In molluscs and men, annelids and albatrosses, gases are exchanged at respiratory surfaces between the atmosphere and the blood (or haemolymph in invertebrates). Gases are then transported between the respiratory surfaces and the tissues by circulation of the blood. The arrangements in insects (and in at least some other terrestrial arthropods) are quite different; gases are exchanged almost directly between respiring cells and the atmosphere (Figure 6.1). Insects possess a system of tracheae and tracheoles — air filled tubes which connect the tissues to the exterior. A small amount of tracheolar fluid is the only barrier between air and cell.

Why is the arrangement of the respiratory system crucial to cold-hardiness? The reason is simple. Because the blood of insects does not have a respiratory function it can be far more viscous than the blood of other animals, thus allowing a more pronounced nutritive role. In all other animals with circulatory systems, the blood viscocity must reflect a compromise between low viscosity

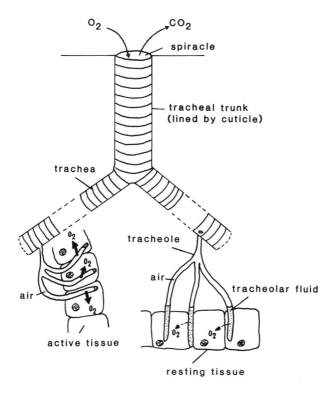

Figure 6.1 Diagram of tracheal respiratory system of insects. Respiration in insects is essentially divorced from the circulatory system and has an automatic feedback system driven by the activity of tissues. Fine tracheoles carry air to the tissues. In resting tissues the blind ends of the tracheoles are filled with tracheolar fluid and the surface area available for gas exchange between tracheoles and tissues is low. When tissues become active they rapidly accumulate waste products in the form of small molecules (e.g. lactate). These molecules are osmotically active, so the intracellular fluids become hyperosmotic to the tracheolar fluid. Water flows out of the tracheoles into the tissues (by osmosis) and a greater surface area between air and tissues becomes available for gas exchange. The lack of involvement of the circulatory system allows insect blood to be much more viscous than in other animals.

(which permits rapid circulation for respiratory purposes) and higher viscocity (which allows the transport of organic molecules, primarily for nutrition, but tends to slow blood flow). In insects the viscosity of the blood is only limited by the capacity of the animal to pump the thick fluid around the body; thus it becomes possible for the blood to contain concentrations of organic molecules orders of magnitude greater than in animals with coupled respiratory and circulatory systems. Some of these organic compounds are involved in cold-hardiness.

Cold resistant ('cold-hardy') insects and other arthropods have been divided into two categories, 'freezing-susceptible' ('freezing intolerant') and 'freezing-tolerant' (Salt, 1961). Freezing-susceptible species die if frozen, but survive to substantial subzero temperatures by supercooling (i.e. by maintenance of body fluids in the liquid state below their solution freezing points). Many insects and mites tolerate temperatures down to −30 to −50°C by this means; currently the extreme example of supercooling is shown by larvae of the dipteran *Rhabdophaga*. These spend the winter in willow galls and are able to supercool to −66°C (Ring, 1981). Freezing-tolerant species survive the extracellular formation of ice in their tissues and normally start to freeze at what are known as 'fairly high' subzero temperatures (−5 to −12°C), so it is clear that they too have some supercooling abilities. Freezing-tolerant species may survive to temperatures as low as the freezing-susceptible arthropods.

The physiology and biochemistry of cold-hardiness in terrestrial arthropods have attracted much research. Many useful reviews have been published in the past two decades including those of Asahina (1969), Block (1982), Duman *et al.* (1982), Miller (1982), Somme (1982), Zachariassen (1982), Baust and Rojas (1985), Cannon and Block (1988) and Bale (1989). The earliest studies demonstrated high levels of the polyhydric alcohol glycerol in cold-hardy insects (e.g. Salt, 1957, 1959, 1961; Somme, 1964, 1967), whether they were freezing-susceptible or freezing-tolerant. Glycerol, other polyols since implicated in cold tolerance (arabitol, erythritol, fucitol, inositol, mannitol, rhamnitol, ribitol, sorbitol, threitol, xylitol), plus sugars such as trehalose and glucose apparently have two functions. In freezing-susceptible animals the polyols are believed to depress the supercooling point (SCP) of body fluids, while in freezing-tolerant species they function as cryoprotectants, inhibiting freezing damage. In general glycerol is synthesized in insects from glycogen (Wyatt, 1967), but it is possible to derive glycerol from triglycerides (by the action of lipases), so fat sources should not be ignored. Concentrations of polyols can be very high; *Rhabdophaga*, the dipteran mentioned above, can contain glycerol at a concentration of more than 20% of the larval wet weight during winter (this species also contains high concentrations of glycogen), while Miller and Werner (1980) reported concentrations of 4–5 molar glycerol in other willow gall insects. These larvae had a haemolymph melting point of −19°C (due to the colligative effects of this high concentration of small molecules) and could supercool to −60°C! Control of polyol production and accumulation has attracted some study. Both desiccation and low temperature are capable of inducing polyol accumulation and it has been suggested that polyol production is probably a response to a suite of stresses which combine to affect the state/quantity of water within a terrestrial arthropod (Baust *et al.*, 1985). In the freezing tolerant larvae of the tephritid fly, *Eurosta solidaginis* glycerol production is stimulated

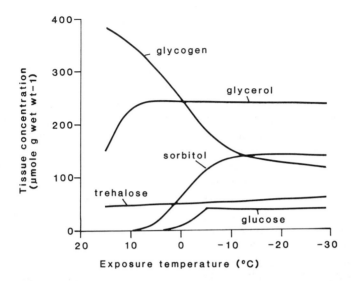

Figure 6.2 The cold induced conversion of glycogen into sugars and sugar alcohols in larvae of the fly *Eurosta solidaginis*. Note that different exposure temperatures favour the production of different cryoprotective agents. Slightly modified from Storey *et al.* (1981).

by sustained field temperatures between 10 and 15°C; below 10°C glycerol accumulation becomes inhibited and between 5 and −5°C sorbitol accumulation takes over (Figure 6.2). Both polyols are derived from glycogen in *Eurosta*; it appears that one of the enzymes controlling carbon flow (phosphofructokinase) acts as a temperature-sensitive gate. At fairly high autumn temperatures the enzyme is inactive and glycolysis yields glycerol (which remains in the haemolymph throughout the winter). As temperatures drop below 5°C, the active form of phosphofructokinase appears and sorbitol is produced instead, either from fructose or glucose (see Franks, 1985 for discussion). Unlike glycerol, sorbitol can be used to resynthesize glycogen. The extent to which polyol production is under neurendocrine control in terrestrial arthropods is as yet unclear.

Differences between freezing-susceptible and freezing-tolerant species, lie not in the possession of different concentrations of polyols, but in the presence or absence of ice nucleating agents. Small samples of pure, dust free water can easily be cooled without freezing to about −20°C. If great care is taken (e.g. by placing samples in extremely clean capillary tubes, cooling slowly and avoiding vibration), supercooling to −40°C is feasible. If solutes are added to water, then the SCP is reduced, and the extent of the reduction in SCP is proportional to the concentration of solute. As a rule of thumb, the depression in SCP is about double the size of the reduction in melting point (Mackenzie, 1977; Block and Young, 1979). Thus, if a distilled water sample supercools to −20°C and melts at 0°C, then a molar solution of mannitol should supercool to about −23.8°C and melt at −1.9°C. Supercooled liquids are metastable (i.e. are in a precarious state); their molecules have cooled and lost kinetic energy, but have not assumed the rigid lattice of the solid. There is no special reason why they should; it is purely a matter of chance whether they spontaneously adopt the pattern necessary for a crystal lattice to form. Shaking or stirring the liquid, or adding dust (nucleating agent) to it may cause it to solidify, but dropping a crystal of its own solid into it is most likely to cause the propagation of the crystal lattice, since it acts as a template or jig for further crystal growth. In the context of terrestrial arthropods two types of nucleating agents are known; material in the gut (whether food particles or ingested 'motes', i.e. dust taken in with food), or organic molecules in the haemolymph which resemble ice nuclei and therefore initiate ice formation. Insects and other terrestrial arthropods that are not cold-hardy (e.g. animals which remain active during winter) have substantial quantities of nucleating agents; they have food in the gut and a variety of proteins and carbohydrates in all body fluid compartments. Accordingly, when cooled to below the melting point of their body fluids, they become vulnerable to freezing — the nucleating agents trigger ice formation within the intestine, haemolymph and intracellular fluids and they die, usually at temperatures above about −6°C, since the presence of nucleating agents reduces supercooling capacity (the difference between SCP and MP) to about 5 deg. C (Krog *et al.*, 1979). Freezing of the gut contents appears to be particularly deleterious as it causes gut swelling and mechanical damage.

Cold-hardy, but freezing-susceptible arthropods reduce the quantity of nucleating agents in their bodies. Primarily they do this by avoiding feeding before entering diapause. Several studies have demonstrated that starved insects have lower SCPs than fed animals, and this is probably a major reason for the difference between the SCPs of active, feeding summer insects (laden with nucleators) and of their dormant winter (starved) phases. Freezing-susceptible arthropods must also reduce the amount of intra- and extracellular nucleators, or they must mask them in some way. Zachariassen *et al.* (1982) demonstrated that haemolymph taken from freezing tolerant beetles (*Eleodes blanchardi*) may be diluted considerably whilst still maintaining its nucleating effect. However, when diluted beyond a factor of about 1000, its effectiveness falls off dramatically (Figure 6.3), suggesting that a certain molecular concentration is needed for a substance to act as a nucleator. Some freezing susceptible arthropods are capable of supercooling to about −40°C, the temperature at which pure, nucleator-free water finally freezes because small, spontaneously developed 'ice embryos' grow to a critical size at which the free energy of the system favours further growth and water freezes ('homogenous nucleation'); this contrasts with 'heterogenous nucleation' which is based on

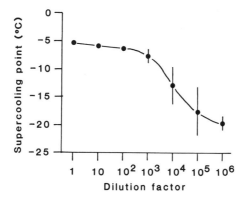

Figure 6.3 Effect of dilution on nucleator activity of haemolymph samples from the beetle *Eleodes blanchardii*. 5 μl samples of haemolymph were repeatedly diluted by 50 μl volumes of 0.9% NaCl solution. Not until the haemolymph is diluted by more than a thousand times does the nucleator activity decrease significantly. Redrawn from Zachariassen (1982).

nucleation by motes or other non-ice nucleation agents. Zachariassen (1985) believes that insects supercooling to such low temperatures do not freeze because of homogenous nucleation, but by the action of weak nucleation agents; he has coined the term 'semihomogenous nucleation' for this process.

Since haemolymph organic nucleators are products of metabolism, starvation is likely to result in reduced concentrations of nucleating agents. It has been suggested that some of the polyols and sugars accumulated in the blood and tissues of freezing-susceptible arthropods act to mask or inactivate nucleators (e.g. Baust and Morrissey, 1975), but *in vitro* experiments by Zachariassen and Hammel (1976) and Lee *et al.* (1981) have demonstrated that glycerol and a variety of other polyols and sugars only act to reduce the SCP by virtue of their osmoconcentration; they are ineffective in the presence of nucleators. In summary, therefore, cold-hardy, freezing-susceptible arthropods survive low temperatures by the possession of large quantities of solutes in the blood (which reduce the SCP), together with low concentrations of potential nucleators. It must be remembered that freezing-susceptible species are in an unstable situation when supercooled. Whether ice nuclei form in the body fluids is largely a probabilistic matter, dependent on duration of exposure to low temperature, the degree of supercooling and the likelihood of nucleator molecules attaining sufficiently high local concentrations to trigger ice formation. Prolonged cold weather is therefore likely to cause high mortality, not necessarily because the environmental temperature finally falls below the blood SCP, but because the degree of supercooling increases instability, and the long period of exposure allows a longer period for nucleation to occur by chance.

Freezing-tolerant arthropods have a different cold hardiness strategy. Although they usually supercool to some extent, their SCP values are high and it appears that they induce freezing of the extracellular fluids by the elaboration of nucleating agents (peptides/proteins/lipoproteins) in the haemolymph. The haemolymph nucleating agents are potent, thus giving preferential freezing in the extracellular fluids, rather than in the cells or gut where less potent nucleators are present according to Zachariassen (1982) (Figure 6.4), though other authors have contended that cells do not contain nucleators, and that cell membranes slow/prevent ice penetration into the supercooled interior (Mazur, 1970; Frankes, 1982). It must be stressed that the elaboration of nucleators is an adaptive, positive process; freezing-tolerant, cold hardy arthropods show higher concentrations of nucleating substances in the winter than in the summer. Because the extracellular fluid begins to freeze at relatively high subzero temperatures, ice formation proceeds slowly enough for water fluxes across cell walls to take place sufficiently quickly to prevent the build up of osmotic gradients and to reduce the probability of intracellular ice formation. As temperatures fall further, freezing 'tolerant' arthropods eventually die. The lower lethal temperatures for various species and the cause(s) of death are diverse.

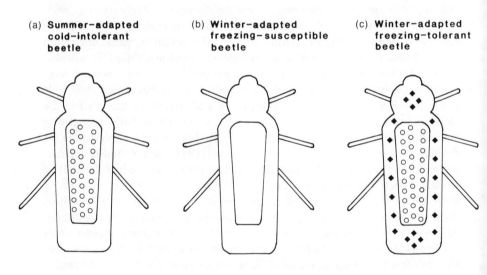

(a) **Summer–adapted cold–intolerant beetle**

(b) **Winter–adapted freezing–susceptible beetle**

(c) **Winter–adapted freezing–tolerant beetle**

o°o nucleating agent in gut or intracellular compartment

♦°♦ nucleating agent in haemolymph

Figure 6.4 Distribution of nucleating agents in the body compartments of beetles. Redrawn from Zachariassen (1982).

Because the freezing process raises the osmotic pressure of the unfrozen part of the extracellular body fluids, it is possible that the cells eventually die of raised internal osmolarity (causing denaturing of enzymes and other proteins) or because of cellular shrinkage (as a result of osmotic outflow of intracellular water). Meryman (1970) was the original proponent of the concept of a 'minimum tolerable cell volume', the minimum volume being determined by the degree to which cell membranes can resist compression. It has been suggested that cold-hardy arthropods may remove lipids from the membranes during shrinkage — thus providing relief from membrane compression.

Alternatively, the falling temperature may eventually cause the phospholipids of membranes to change phase. It is generally held that cell membranes need to be fluid to function. Phospholipid membranes are complex structures with embedded proteins, so phase transition is not likely to be abrupt. However, if they start to solidify, a range of membrane functions are likely to fail. It is in preventing these secondary effects of ice formation that polyols act as cryoprotective agents. As pointed out by Cannon and Block (1988) they can stabilize proteins (including enzymes) against cold denaturation (as originally demonstrated by Hochachka and Somero, 1973), they can inhibit lipid phase transitions, and they can also reduce osmotic dehydration, apparently by locking up water in unfreezable form ('bound water'). However, the concept of unfreezable, bound water is regarded with deep suspicion by some scientists. Franks (1985) believes that the only unfreezable water present in animals is that existing in pores/capillaries (of approx. 1 μ diameter or less) penetrating solid substrata.

A number of freeze-tolerant insects can survive to temperatures well below −25°C. It has been difficult to see how intracellular freezing or destructive osmotic changes could be avoided at such low temperatures. Recently an attractive hypothesis has been presented to explain such survival (see personal communications of Wasylyk and Baust in Bale, 1989). It would seem that at these low temperatures, in multicomponent cryoprotective systems (made-up of glycerol, sorbitol, fructose and trehalose), the larvae of the fly *Eurosta solidaginis* exhibit vitrification of unfrozen water, both within and outside cells. Vitrification means that the water enters a glassy state in which it is noncrystalline and extremely viscous (virtually solid); molecular diffusion essentially ceases. Once vitrification takes place, osmotic flow of water (and hence cell shrinkage) will cease. On rewarming it would be necessary for the glass to change state to the liquid form of water without intermediate ice formation; this seems to be achieved.

As research on cold-hardy terrestrial arthropods has progressed, great diversity in response has been revealed, and to some extent the nature and basis of cold-hardiness has become confused (see the short review of Baust and Rojas, 1985 for a stimulating critique). Broadly speaking, freezing-susceptible supercooling species are more common than freezing-tolerant species (this is true both of high northern latitudes and the Antarctic), but there are some species which cannot be so easily categorized. For example, the larva of the beetle *Pytho deplanetus* which lives at high altitude (2000+ m) in the Rockies supercools to −54°C, and this is its normal response to cold. However, on occasion (particularly if it contacts ice whilst in its hibernaculum) it will freeze — and can then tolerate freezing to −55°C because glycerol protects against cell damage. In 1984 Duman demonstrated that larvae of the Cucujus beetle (*Cucujus claviceps*) were capable of changing their overwintering mechanism between freezing susceptibility and freezing tolerance. More receptly still Kukal and Duman (1989) have found that an apparent switch in strategy (from freezing tolerance to freezing susceptibility) which occurred in two beetle species (*Cucujus claviceps* between 1979 and 1983 and *Dendroides canadensis* between 1980 and 1981) in Indiana, was real and not due to northerly migrations of low latitude freezing susceptible animals. It is therefore becoming clear that the categories 'freezing-tolerant' and 'freezing-susceptible' are leaky rather than watertight! The triggers for such switches in strategy are presumably environmental, but Kukal and Duman were unable to disentangle the factors responsible in the case of the beetle species.

Some arthropod species supercool considerably, yet do not have high concentrations of polyols (they probably survive by efficient scavenging of nucleators from the haemolymph), while in contrast it has become clear that many species have multi-component systems in which several cryoprotective agents are involved. Multiplicity of cryoprotective agents seems to be adaptive, since it avoids any single component reaching toxic concentrations. As Baust and Rojas point out, early workers often described single component systems (particularly those based on glycerol) which have been unstudied since. Some of these would probably be shown to be multicomponent systems if modern techniques were applied. Finally, it must be remembered that the polyols and sugars ('low molecular weight antifreezes' — Duman *et al.*, 1982) may have functions other than cryoprotection — they tend to reduce water loss and protect against desiccation damage (vital in inactive insects which are not eating or drinking); they may also be important energy stores.

A category of arthropod cryoprotective agents not mentioned so far in this section was first discovered by Ramsay (1964) in the larvae of the mealworm beetle *Tenebrio molitor*, although the importance to freezing resistance was first revealed by Duman (1977a,b) who investigated another tenebrionid beetle, *Meracantha contracta*. This category encompasses a range of 'high molecular weight antifreezes', better known as 'thermal hysteresis proteins' (THPs) or 'antifreeze proteins'. THPs are macromolecules of molecular masses between 9000 and 17 000 (Duman and Horwath, 1983). First found in beetles, they are now known from many species of overwintering insects and spiders, though not recorded yet from mites; they have been found in all freezing-susceptible hibernating insects studied so far (Zachariassen, 1985), but are also known from freezing-tolerant beetles (Duman, 1979). These macromolecules lower the freezing point of arthropod body fluids, thus creating a difference (hysteresis) between melting point and freezing point. They function in similar fashion to the antifreeze glycoproteins/glycopeptides of teleost fish (section 6.2), since they effectively inhibit growth of ice crystals by increasing the radius of curvature of the ice front (Chapter 7). However, unlike the fish antifreezes, THPs do not contain a sugar component. Several species of beetle appear to rely solely upon THPs for cold-hardiness, particularly species such as *Meracantha contracta* which live in damp habitats where cross-integument inoculation by ice crystals is an environmental hazard. Generally speaking, THP production, like polyol accumulation, is a seasonal phenomenon, with concentrations and corresponding degrees of thermal hysteresis being greatest in winter. Several species combine polyol and THP elaboration to produce profound depression of both freezing and supercooling points. The beetle *Uloma impressa* is a good example. In summer (June) the haemolymph melts at −0.88°C and freezes at −2.00°C (thermal hysteresis 1.12 deg. C; the insect's supercooling point is −6°C). In winter the beetle overwinters as an adult under loose bark on dead trees. Both THPs and glycerol are accumulated, the haemolymph melting point falls to −9.90°C because of the high concentration of glycerol and the freezing point drops to −14.68°C. The thermal hysteresis of 4.76 deg. C is due to heightened THP levels. The supercooling point of the beetle in winter is −21°C. The fall in supercooling point is probably mainly due to the presence of glycerol, but there is some suggestion (so far unsubstantiated) that THPs may also cause supercooling point depression (Duman *et al.*, 1982). Several THP molecules from insects and fish have been analysed and it is clear that there are inter- and intraspecific differences. The mealworm *Tenebrio* possesses four different THPs, each with different hysteresis properties; it has been suggested that they may act in synergistic fashion when mixed (Schneppenheim and Theede, 1980). As with polyols, there is good evidence that THPs may have functions other than freezing point depression. The mealworm *Tenebrio molitor* is well known for its ability to reabsorb water from the rectal complex; it also takes up water vapour from the atmosphere. THP molecules have been implicated in this process (Ramsay, 1964; Grimstone *et al.*, 1968). THPs may also act as useful sources of amino acids for protein building at the end of winter.

Baust and Rojas (1985) made some fundamental criticisms of experimental approach in studies of insect cold hardiness. They pointed out that most workers had subscribed to the view of Salt (1966) that supercooling points should be established by using a standard cooling rate (1 deg. C^{-1}) to permit comparability; there was also an underlying set of untested assumptions about supercooling points — notably that animals survived all temperatures above the supercooling point — with little consideration given to duration of exposure. Baust and Rojas gave evidence from the work of themselves and others which demonstrated that slower cooling rates (more environmentally realistic) could allow survival to lower temperatures. The author has sympathy for these criticisms, having found a similar divorce from environmental reality in the laboratory studies of salinity, temperature and pollutant responses of fish and invertebrates (Davenport, 1982). It is all too easy to sacrifice effective environmental simulation to laboratory convenience and (often spurious) comparability! More recent investigations and reviews (e.g. Bale, 1989) have taken on board these criticisms and stimulated the development of more comprehensive, integrated studies.

The importance of considering environmental influences on cold-hardiness in terrestrial arthropods is nowhere better demonstrated than in the consideration of external sources of ice nuclei. Most ice nuclei present in the atmosphere are minute clay-silicate particles ($0.1–1.0\mu$ in diameter) which are effective heterogenous nucleators at temperatures below about $-15°C$ (Mason, 1971). The amount present in the atmosphere differs on a day to day basis as will the likelihood of such particles being ingested by, or falling upon, terrestrial arthropods. Sudden increases in concentration of these particles have been likened to storms (Pruppacher and Klett, 1978). Nucleating agents may even be living; several bacterial strains are now known to function as highly effective nucleators, with *Pseudomonas syringae* producing active nuclei at temperatures as high as $-4°C$ (Lindow, 1983). There is some evidence that Antarctic mites may lose supercooling ability as a result of feeding upon nucleating microbes (see Cannon and Block, 1988).

6.2 FREEZING-TOLERANCE IN TERRESTRIAL VERTEBRATES

Until a few years ago it was generally accepted that no vertebrate survived freezing (beyond a certain amount of freezing of superficial tissues, which in any case caused damage and necrosis). However, in 1982 Schmid reported freezing-tolerance from three species of anuran amphibian from cold temperate areas of the USA, the wood frog *Rana sylvatica*, the grey tree frog *Hyla versicolor*, and the spring peeper, *Hyla crucifer*. All overwintered at the soil surface where they would be covered with leaf litter or snow, but where temperatures fell to -2 to $-8°C$. The frogs all appeared to freeze solid; measurements indicated that 35–48% of body water was frozen, but only in the extracellular compartment. Heart rate and breathing are completely abolished during freezing, but return over a few hours on thawing. Subsequent experiments (Storey and Storey, 1984, 1986; Storey, 1985) have established that frogs begin to freeze at about $-2°C$ (Figure 6.5), being supercooled by about 1.5 deg. This limited supercooling ability probably indicates that the frogs do not need to synthesize nucleating agents.

It has been found that frogs can survive freezing for several days, with

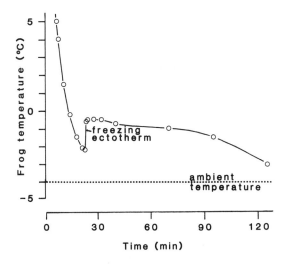

Figure 6.5 Cooling curve for a male adult spring peeper [a tree frog] *Hyla crucifer* held in an incubator at −4°C. This curve demonstrates that spring peepers held at subzero temperatures initially undercool but then allow freezing of the body fluids (positively shown by the freezing ectotherm. Redrawn from Storey (1985).

mature animals tolerating longer periods (up to 13 days in *Hyla versicolor*) than immature frogs. All species produce cryoprotectants which apparently inhibit intracellular freezing. Adult *Hyla versicolor* accumulate glycerol (0.3 molar), but all of the other species use glucose (immature *Hyla versicolor* employ a mixture of glucose and glycerol). Glucose accumulation has been studied in detail. In contrast to insects, no accumulation of cryoprotectants in the blood prior to exposure to freezing has been demonstrated, though glycogen and lipid are accumulated in the autumn months. Instead, glucose synthesis and accumulation is triggered by exposure to freezing conditions in the 0 to −2°C range (Figure 6.6). In *Rana sylvatica* normal blood glucose levels were around 3 mmole l^{-1}, but rose to as much as 325 mmole l^{-1} in frozen frogs (Storey and Storey, 1984). Glucose appears to be synthesized only in the liver, since freezing caused a marked decrease in liver glycogen, but other sources of glycogen (e.g. in heart and kidney) were unaffected. Liver activity levels of enzymes needed for glucose synthesis (e.g. phosphorylase and glucose-6-phosphatase) also rose dramatically (1.4–5.2 fold). Effectively the frogs become diabetic for the duration of the freezing episode; they recover normal glucose and enzyme levels within about 10 days after thawing. The high blood glucose levels during freezing have several implications, as yet unresolved. High blood sugar levels have been reported to cause problems for protein biochemistry in other

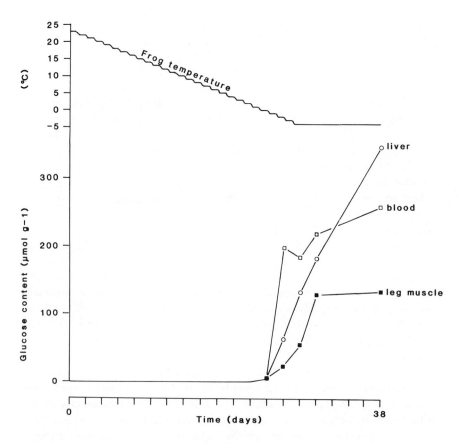

Figure 6.6 Levels of glucose in the blood, liver and leg muscle of frogs (*Rana sylvatica*) exposed to decreasing temperatures (culminating in exposure to −4°C for 11 days). Drawn from the data of Storey and Storey (1984).

vertebrates, and there is probably an interesting story as far as hormonal control of glucose metabolism by insulin and glucagon is concerned.

Storey and Storey (1984) also demonstrated that all tissues accumulated lactic acid and lost ATP/creatine phosphate during the period of freezing. This indicates a reliance on anaerobic glycolysis, which confirms that the tissues are isolated and ischaemic, by virtue of the shut down of blood circulation and lung ventilation (much of frog respiration is cutaneous, but the ischaemic skin will presumably inhibit this route of oxygen uptake too). Storey and Storey suggest that the limitation on survival time in the frozen state is probably linked to the period for which tissues can rely on stored energy. Possibly the brain tissue is rate-limiting in this respect, since it is a general feature of vertebrate brains that they rely heavily on energy supplied from elsewhere in the body.

During the past 2–3 years another vertebrate, the painted turtle *Chrysemys picta*, has attracted study. The painted turtle is a fairly small freshwater emydid turtle, closely related to the red-eared slider (*Trachemys scripta elegans*) sold in pet shops throughout the world. Painted turtles live in cool temperate areas of the USA (e.g. Michigan, Minnesota, Nebraska). They lay eggs in late spring/early summer in shallow nests (7–12 cm deep). The eggs hatch in late summer or early autumn, but most of the hatchlings remain in the nest throughout the subsequent winter, only emerging in early spring. Packard *et al.* (1989) showed that nest temperatures in the field may fall as low as $-6.2°C$ during the severe winters of Nebraska. Pauktis *et al.* (1989) exposed painted turtles to a cooling cycle (0 to -8 to $0°C$) over a 29 h period in the laboratory and found that the turtles survived by supercooling. Further experiments demonstrated supercooling to $-8.9°C$ (at which point the turtles spontaneously froze). Even when freezing took place, some individuals survived, one turtle remaining alive after freezing and further cooling to $-10.4°C$. These data suggest that painted turtles have a freezing avoidance strategy based on supercooling, and only rely on freezing tolerance *in extremis* (by means as yet unknown). However, Packard *et al.* (1989) point out that supercooling under field conditions may be difficult as the hatchlings are packed together in damp earth — ideal conditions for ice nucleation.

At present few ectothermic vertebrates are known to tolerate freezing of their body fluids. However, it seems likely that more examples will be discovered, particularly amongst those groups of amphibia and reptiles which live at high latitude and spend the winter in some form of hibernation.

7 Subzero temperatures and marine ectotherms

7.1 INTRODUCTION

Sea water (osmolarity approximately 1000 mOsmoles 1^{-1}) freezes at about
$-1.9°C$. Most marine invertebrates have body fluids which are iso-osmotic
with sea water and so have a similar freezing point. Provided that they are not
in prolonged contact with ice crystals, such animals are exposed to little risk of
freezing whilst immersed in sea water. Even so, some antarctic and arctic
amphipods maintain slightly heightened blood concentrations, which reduce
the risk still further (Rakusa-Suszczewski and McWhinnie, 1976; Davenport,
1981).

The situation for fish is rather different. While elasmobranchs (sharks,
skates and rays) have high blood urea concentrations which raise blood
osmolarity to levels comparable with invertebrate body fluids, the blood of
teleost fish is more dilute (reflecting their freshwater/brackish water ancestry).
Most marine teleost plasma samples have osmolarities between 300 and 500
mOsmoles 1^{-1} and might therefore be expected to freeze at -0.6 to $-1.0°C$. In
areas of the Arctic and Southern Oceans where open seawater temperatures
between 0 and $-1.9°C$ are common, especially in winter, such fish are poten-
tially vulnerable to freezing.

A small proportion of marine ectotherms exploit the intertidal zone where
they may be exposed to much more extreme temperatures when the tide is out.
At very high latitudes the intertidal zone is either permanently frozen or is
barren and lifeless because of seasonal scouring by ice. However, at lesser
latitudes air temperatures of -10 down to $-20°C$ occur regularly during the
colder months of the year. There is a general tendency for more mobile intertidal species (e.g. fish, crabs) either to leave the shore in winter (Chapter 3), or
simply not to be found at high latitude. Thus the shore crab *Carcinus maenas*

moves offshore in the UK during the winter (when air temperatures during night time low tides often drop to $-5°C$ and in severe winters fall below $-10°C$), but becomes less common along the eastern Norwegian coast as one moves northwards, progressively encountering longer and colder winters. Intertidal fish such as the blenny *Lipophrys pholis* display similar winter migrations and distributional characteristics. On the other hand, many sessile or sedentary species (e.g. barnacles, bivalves and gastropods) cannot avoid exposure to freezing temperatures yet are found on the shores of cold and temperate countries, so have to endure them.

7.2 SUPERCOOLING IN DEEP-WATER FISH

A strategy adopted by a number of cold-water fish to avoid freezing is to migrate into deep, cold water during the winter months or to spend their whole life in this environment. Deep water in the arctic has a temperature of $-1.8°C$, but, being slightly warmer than the freezing point of sea water, does not contain ice crystals. Scholander *et al.* (1957) were the first workers to note that fish could exploit deep water as a refuge in which they survived in a supercooled state. They found that fish living permanently in deep water, such as *Boreogadus saida* and *Liparis turneri* had slightly lower blood freezing points (-0.9 to $-1.0°C$) than temperate species but not to an extent which significantly reduced the risk of freezing. They found that if ice was introduced into the environment of these fish, the fish instantly froze, even though the sea water itself did not freeze. In an experimental situation (Figure 7.1), such fish could be held at $-1.5°C$ in an aquarium of sea water. If an ice cube was placed in the water the fish would freeze instantly, presumably because of migration of seeding ice nuclei from cube to fish. If a frozen fish was allowed to partially thaw (so that the outer tissues were unfrozen) and then placed in a smaller aquarium with a live fish, the latter would freeze within a few seconds, confirming the easy migration of ice through peripheral tissues. Clearly such fish are in an unstable, supercooled state, only tenable because of the lack of ice nuclei in their environment.

Since these early observations, it has been recognized that deep water supercooling is a quite common phenomenon, both in northern and southern cold waters. Olla (1977) showed that the cunner (*Tautogolabrus adspersus*), which is an inshore fish of Nova Scotia, overwinters in a dormant, supercooled state beneath rock overhangs at a depth of 15–30 metres; if touched with ice (taken down by divers) they freeze and die. The Antarctic liparid *Liparis devriesii* and zoarcid *Rhigophila derborni* also live in a supercooled state at depth (DeVries, 1974), though the latter species returns to the surface in summer.

The author (Davenport, unpublished data) found that adult capelin, *Mallotus villosus* freeze at $-1.4°C$ in the presence of ice, yet spend their whole life history

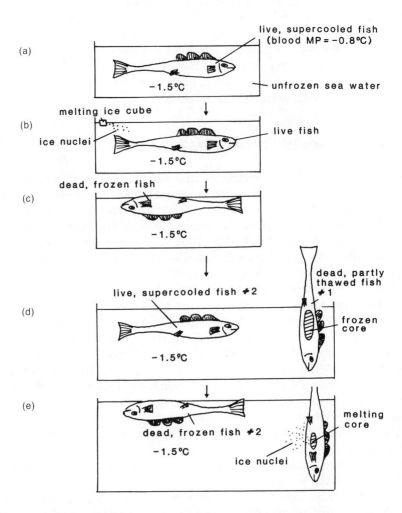

Figure 7.1 Classic Scholander experiment to demonstrate supercooling in fjord cod (*Boreogadus saida*). In (a) a fish is held in an aquarium of sea water at $-1.5°C$; the fish body fluids are supercooled, but the animal can be held in this state for long periods in the absence of ice nuclei. In (b) an ice cube is placed in the aquarium. This melts because the sea water is warmer than the melting point of sea water ($-1.9°C$). In (c) the fish freezes and dies because ice nuclei from the melting cube inoculate the supercooled body fluids. In (d) and (e) the experiment is repeated, but with a partly thawed dead fish instead of an ice cube. Again the live, supercooled fish is killed by inoculation with ice nuclei. From data of Scholander *et al.* (1957).

at high latitudes. There have been reports of capelin freezing as they have been trawled up through icy surface waters (Icelandic fishermen have occasional problems with this phenomenon in the Greenland Strait, since the frozen fish cannot easily be pumped from purse seine nets), and some mass mortalities of capelin trapped in shallow bays during winter have been attributed to freezing (Templeman, 1948, 1965) so it is likely that the large concentrations of capelin in the Arctic Ocean survive in winter by avoiding shallow water containing ice crystals.

7.3 HEIGHTENED PLASMA OSMOLARITY IN COLD-WATER FISH

Some fish do dwell in surface waters even when winter seawater temperatures are as low as $-1.7°C$. Sculpins (*Myoxocephalus scorpius*) and fjord cod (*Gadus ogac*) collected in the summer have a blood freezing point of $-0.8°C$, but in winter the freezing point falls to $-1.5°C$. There may be some 'antifreeze' involvement in this change in blood FP (see below), but much of the change (40–70%, depending upon species) is due to increased plasma concentrations of sodium and chloride, and plasma concentrations may almost double between summer and winter. Two hypotheses appear feasible to explain the changes in blood ionic concentration. Vernberg and Silverthorn (1979) suggested that the fishes' active ion extrusion mechanisms became less efficient at very low temperatures, so that rising plasma concentrations in winter were simply the result of passive inward diffusion of ions as the fish became incapable of sustaining ionic gradients at the summer levels. This is the generally accepted view, but it should be noted that Baltic fish, living in dilute brackish water with a lower ionic content than the fishes' plasma, also show increased plasma concentrations at low temperature, suggesting that the increase is an actively-controlled regulatory process (Oikari, 1975). Whatever the mechanism, the fish are still slightly supercooled (by about 0.2 deg. C) but probably risk freezing only if in prolonged contact with ice.

In the coastal waters of the Southern Ocean, temperatures vary little, always being quite close to freezing. Although much of the survival of Antarctic fish is due to antifreezes (section 7.4), it should be pointed out that the blood osmolarities of these fish are uniformly high (around 600 mOsm Kg^{-1}), nearly twice the concentration of plaice and cod blood.

7.4 ANTIFREEZES IN HIGH LATITUDE FISH

In 1969 DeVries and Wohlschlag reported exciting findings from their studies on Antarctic fish, some of which live in intimate contact with ice platforms, tunnels and shelves and are surrounded by seawater loaded with ice crystals.

They found that such fish had plasma which demonstrated a phenomenon now generally known as thermal hysteresis — it did not freeze (i.e. ice crystals would not propagate in it) until about $-2.0°C$, but on rewarming, melted at about $-0.8°C$, indicating that unfrozen plasma contained some non-colligative agent(s) which reduced its freezing point. Dialysis of plasma revealed that dialysed plasma (i.e. plasma from which macromolecules had been removed) froze and melted at the same temperature (around $-0.8°C$), and it was found

Table 7.1   Occurrence of macromolecular antifreezes in fish. ARC = Arctic; NT = north temperate; ANT = Antarctic; AFP = peptide antifreeze; AFGP = glycopeptide antifreeze

Species	Distribution	Antifreeze type	Molecular weight
Gadiformes			
Gadidae			
Boreogadus saida (arctic cod)	ARC	AFGP	2600–33700
Eleginus gracilis (saffron cod)	NT	AFGP	5000
Gadus ogac (Greenland cod)	ARC	AFGP	2600—33700
Perciformes			
Notctheniidae			
Dissostichus mawsoni	ANT	AFGP	2600–33700
Notothenia gibberifrons	ANT	AFGP	2600–33700
Channichthyidae			
Chaenocephalus aceratus (icefish)	ANT	AFGP	62000
Chionodraco hamatus	ANT	AFGP	10000–39000
Cottidae			
Hemipterus americus (sea raven)	NT	AFP	14000
Myoxocephalus verrucosus (Bering Sea sculpin)	ARC	AFP	5000
Zoarcidae			
Rhigophila dearborni (Antarctic eel pout)	ANT	AFP★	5000
Lycodes polaris (Arctic eel pout)	ANT	AFP★	5000
Pleuronectidae			
Pseudopleuronectes americanus (winter flounder)	NT	AFP	3000–8000

★ These peptide antifreezes are very similar (80% homology of residues) despite great geographical separation

Simplified from Clarke, 1983, with additional information from Schrag *et al.*, 1987.

that the macromolecular fraction of the plasma contained a glycopeptide 'antifreeze'.

DeVries and his coworkers have since extended these studies considerably, but before discussing their work, it ought perhaps to be recorded that Scholander and his coworkers (e.g. Gordon *et al.*, 1962) demonstrated the existence of antifreeze in the northern (Canadian) species, *Myoxocephalus scorpius* and *Gadus ogac* several years earlier, but were unable to quantify its importance or to identify its chemical nature.

DeVries and his group have now studied many fish species from the Antarctic and Arctic (see DeVries, 1980, 1982; DeVries and Eastman, 1982; Schrag *et al.*, 1987; Ahlgren *et al.*, 1988 for review) and it is clear that teleost antifreezes are of widespread cold water occurrence (Table 7.1) and of crucial importance to survival in polar waters affected by ice (see Figure 2.6 for an indication of the

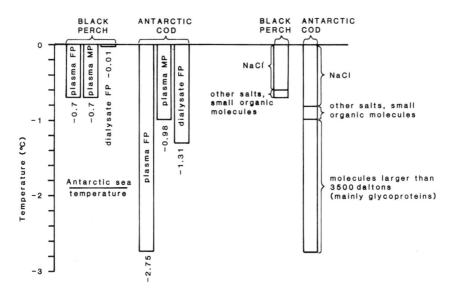

Figure 7.2 Diagram to illustrate importance of antifreezes in Antarctic 'cod' (*Trematomus borchgrevinki*). A temperate species, the black perch (*Embiotoca jacksoni*) is used for comparison. At the left of the figure it may be seen that the freezing point of Antarctic cod plasma (−2.75°C) is much lower than that of the perch (−0.7°C). However, when frozen plasma is melted there is a far smaller difference between the species (−0.7 vs − 0.98°C). Dialysed Antarctic cod plasma (i.e. plasma in which organic molecules have been removed) freezes at −1.31°C, little lower than the whole plasma melting point. This indicates that the organic ('antifreeze') fraction is responsible for the low freezing point in Antarctic fish — as is indicated by the histograms on the right of the figure which display the importance of various plasma constituents to plasma freezing point depression. Redrawn from DeVries (1980).

area of ocean so affected). The antifreezes found so far fall into two categories, glycopeptides and peptides, and in both cases they make up only 3–4% (weight/volume) of the blood and some other body fluids; they can therefore make no significant colligative osmotic contribution to freezing point depression. Dialysis has been the main tool for the detection of antifreezes. Figure 7.2 illustrates the results of dialysis performed upon the blood plasma of an antarctic fish (*Trematomus borchgrevinki*, which possesses antifreeze) and of the black perch, *Embiotoca jacksoni*, which lives in temperate southern waters (DeVries, 1980). When dialysed against distilled water, plasma ions diffused across the dialysis membrane, leaving behind a fluid containing colloidal macromolecules (with a molecular weight greater than 3500 daltons), too large to pass through the pores of the membrane. Whereas the dialysed plasma of the antarctic species still exhibited a considerable freezing point depression despite the total loss of ions (reflecting the presence of the antifreeze), that of the temperate fish did not.

Isolation and purification of the colloidal macromolecules have now been carried out for several species. Glycopeptides in antarctic fish have been found to consist of repeating units of glycotripeptides (Figure 7.3), which are themselves made up of alanine-alanine-threonine amino acid sequences, with a polar (hydrophilic) disaccharide (galactose-N-acetylgalactosamine) attached to each threonine molecule. The separation distance between polar disaccharides is 7.36Å, which corresponds to the distance between alternate oxygens along the c-axis of the ice lattice (Fletcher, 1970). Glycopeptides in arctic fish

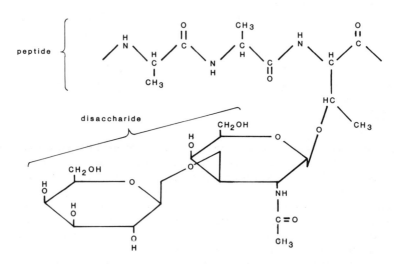

Figure 7.3 Basic repeating unit of molecules of glycopeptide antifreezes. Redrawn from DeVries (1980).

Table 7.2 Molecular weights of glycopeptide antifreezes isolated from the blood plasma of *Pagothenia (Trematomus) borchgrevinki*

Glycopeptide fraction	Molecular weight
1	33 700
2	28 800
3	21 500
4	17 000
5	10 500
6	7900
7	3500
8	2600

From DeVries and Lin, 1977.

are generally very similar to those found in antarctic fish, although in *Eleginus gracilis* threonine is replaced by arginine (O'Grady *et al.*, 1982). DeVries and Lin (1977a) found that there were eight glycoprotein fractions (separated electrophoretically) in antarctic *Trematomus borchgrevinki*, each with a different molecular weight (Table 7.2); the same eight glycoproteins were found in *Gadus ogac* from Labrador (Van Voorhies *et al.*, 1978), strongly suggesting a repeated convergent evolution of antifreezes in unrelated groups of fish separated by geographical barriers. A rather different glycoprotein composition is found in *Eleginus gracilis* (the saffron cod of the Bering Sea). As well as the difference in amino acid sequence mentioned above, the freezing point depression per unit weight is reduced by comparison with that of the glycoproteins already discussed, and only three glycoprotein fractions (averaging 5000 daltons in molecular weight) are present.

Peptide antifreezes were originally thought to occur only in arctic and north-temperate fishes. Duman and DeVries (1976) isolated three peptides from the blood of the winter flounder, *Pleuronectes americanus*, which encounters temperatures of −1.4°C in the winter off eastern Canada and the northeastern USA, yet does not freeze even when in contact with ice. The species only exhibits significant peptide concentrations in winter. The peptides, which are apparently of α-helical form, contain only eight amino acids, of which alanine makes up two-thirds of the residues. Most of the rest of the peptide chains are made up of the amino acids threonine and aspartic acid, which are polar, the former having a hydroxyl group, the latter a COO⁻ group. DeVries and Lin (1977b) found that the polar amino acids are in clusters separated by long alanine sequences, the separation distance being 4.5 Å, which is also the repeat distance of oxygens in the ice lattice along the α-axis of an ice crystal. The closely related *Pleuronectes quadrituberculatus* (Alaskan plaice) has a very similar peptide antifreeze, but that of the Bering sculpin (*Myoxocephalus verrucosa*) is different, having six electrophoretic variants and rather more amino acids

(Raymond *et al.*, 1975). Schrag *et al.* (1987) recently reported upon peptide antifreezes found in fish (eel pouts) of the family Zoarcidae. Members of this family are found at high latitudes in both hemispheres and Schrag *et al.* found that the primary structures of the antifreeze peptides of the antarctic eel pout (*Rhigophila dearborni*) and the arctic eel pout (*Lycodes polaris*) were very similar ('high degree of homology') with 80% of the structures being identical despite the wide geographical separation. The zoarcid antifreeze peptides are also interesting because their structure is different from other peptide antifreezes and there are no regular arrangements of polar groups, making it difficult to see how the mode of action can be similar to that of other antifreeze peptides.

The mode of action of all antifreezes is somewhat obscure. If a 1% solution of purified glycoprotein antifreeze solution is held at a temperature between its freezing and melting points and an ice crystal is introduced, then that ice crystal does not show detectable growth, even after a period of a week (DeVries, 1976). This observation was confirmed for 2% solutions of glycopeptide and peptide antifreezes over a period of 10 days (DeVries, and Lin, 1977a; Raymond and DeVries, 1977), and indicates that the antifreeze molecules do not simply slow ice crystal growth; they totally inhibit it. There is considerable evidence that antifreeze molecules bind to ice, and that the polar components (disaccharides in glycopeptide antifreezes; threonine and aspartic acid in non-zoarcid peptide antifreezes) are involved in the binding process. Chemical alteration of the polar groups results in loss of antifreeze activity (see DeVries, 1980, 1982 for review). Models proposed by DeVries (Figure 7.4) suggest that hydrogen bonding is involved in this 'adsorption-inhibition' mechanism, with glycopeptides binding parallel with the c-axis of the ice lattice and peptides binding parallel with the α-axis. DeVries also suggests that the presence of adsorbed macromolecules on the surface of ice crystals effectively obscures parts of the ice lattice, so that any water molecules joining the lattice are forced to join in intermediate areas, resulting in growing zones having high radii of curvature (Figure 7.4) rather than forming part of a broad growing front. It is a feature of ice crystal growth that microcrystalline high radius crystal growth requires a lower temperature than growth of broader fronts, so any mechanism forcing microcrystalline ice formation would automatically lower the effective freezing point of plasma. There is some evidence for this hypothesis; ice crystal growth is qualitatively very different in solutions of antifreeze than in solutions containing salts alone. In antifreeze solutions, ice forms very rapidly in the form of fine sharp needles 10–20 μm in diameter (Lin *et al.*, 1976; Raymond and DeVries, 1977), rather than slowly in the shape of plates or leaves.

Initial studies of fish antifreezes concentrated upon plasma levels of the macromolecules (primarily because of ease of study), but of course it is necessary for almost all extracellular fluid compartments of the body to contain antifreeze if freezing is to be avoided. Antifreezes have been detected in peritoneal fluid and in interstitial fluids from organs such as the heart, liver and

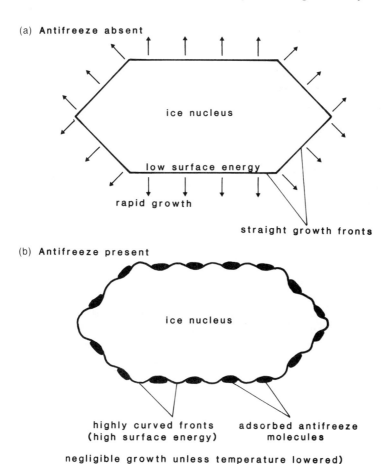

Figure 7.4 Diagram illustrating adsorption-inhibition model of antifreeze action. Glycopeptides adsorbed onto the surface of the ice crystal cause greatly increased curvature of the growing fronts of the crystal. This increased curvature causes high surface energy; growth can only continue if the temperature of the system is lowered. Simplified from DeVries (1980).

locomotory muscles. No antifreezes have been found in the interstitial spaces of the brain, but this organ is surrounded by extradural fluid which does contain glycopeptide antifreeze and presumably acts as a barrier to ice migration. Not all of the different sizes of macromolecules are present in all fluids; Ahlgren *et al.* (1988) pointed out that small sized molecules alone were found in interstitial fluids whereas all sizes were found in extracellular fluid. This suggests that there is some correlation between molecular size and function, but explanation of the requirement for as many as eight differently sized macromolecules (instead of a single one) in polar fish is not yet forthcoming. It should also be borne in mind that some of these molecules may have functions beyond cryoprotection. It is not clear whether antifreeze glycopeptides are found intracellularly. Early publications (e.g. DeVries, 1976) indicated that they were, but the work of Ahlgren *et al.* showed that labelled glycopeptides did not pass from plasma to the intracellular compartment. This does not exclude the possibility that antifreeze glycopeptides are synthesized intra-cellularly, but the available evidence indicates that synthesis takes place in the liver (O'Grady *et al.* 1982), and that the antifreezes are carried passively around the body with plasma proteins.

No antifreeze has been found in the urine of any of the fish featuring these macromolecules. In the case of the antarctic fish studied, this is not too sur-prising, as they all feature aglomerular kidneys (Dobbs and Devries, 1975a, 1975b), that is kidneys in which urine is formed not by ultrafiltration through a porous membrane (which might be expected to allow the passage of the smaller antifreeze macromolecules), but by cellular secretion into the lumen of the nephron. The situation for northern polar fish with peptide antifreezes in their plasma is rather different as they possess glomerular kidneys in which ultrafiltration takes place, and through which molecules as large as inulin (5000 daltons) freely pass. Petzel and DeVries (1979) report that pores in the endo-thelial lining of the glomerular capillaries are lined with negatively charged proteins, which repel the negatively charged peptide antifreeze molecules ('charge repulsion mechanism') and therefore prevent them entering the filter-ing pores. One southern zoarcid species, *Rhigophila dearborni*, the Antarctic eel pout, possesses both a glomerular kidney and an acidic peptide antifreeze (which would not be repelled by a negatively-charged epithelial pore). In this case the glomeruli are few, and are unable to filter inulin or antifreeze, apparently because of thickening of the filtration barrier (DeVries, 1980).

All of these structural features obviously result in the conservation of anti-freeze within the body of the fish. Turnover measurements suggest that antifreeze molecules have a lifetime of about a month in polar fish (see Clarke, 1983 for review), so it would seem that renal conservation of antifreeze ensures that the possession of antifreeze does not require great expenditure of energy. However, the other consequence of renal conservation is that the urine (isos-motic with blood plasma) contains no antifreeze, so is presumably vulnerable to freezing. Why it does not do so is as yet obscure, but Ahlgren *et al.* (1988) report that the urethra has a tight sphincter guarding its outlet, and also contains mucus.

It has also been found that fluids within the eye contain antifreeze molecules, but only in trace quantities (1–2 mg ml^{-1}). As with urine, the aqueous and vitreous humours are secreted, not filtered fluids, so are effectively salt solutions. Ahlgren *et al.* (1988) have shown that the fluids within the eye have a melting point of about $-1.2°C$, so are undercooled by about 0.7 deg. C in

Antarctic sea water at $-1.9°C$. Turner *et al.* (1985) have shown that nucleation by ice crystals is prevented by the transparent head skin and spectacle which overlie the cornea; both are fortified with antifreeze glycopeptides.

Another fluid which might be thought liable to freeze is that within the gut. Marine teleosts have to drink sea water to replace water lost by osmotic flow from the body fluids to the hyperosmotic outside medium (see Rankin and Davenport, 1981; Davenport, 1985a for discussion). In polar regions they will also ingest ice crystals. In the stomach, the osmolarity of ingested fluid remains high, but as salts are pumped out of the gut fluid along the intestine (a procedure necessary to drive the osmotic uptake of water), the gut fluid becomes progressively more dilute and therefore more vulnerable to freezing. Experiments with radiotracer-labelled glycopeptides have shown that plasma antifreezes reach the gut lumen, being secreted by the gall bladder. The bile duct of fish, as in other vertebrates, delivers bile into the small intestine immediately after the stomach, so antifreeze is present in all of the vulnerable parts of the gut. Glycopeptides are not reabsorbed (or hydrolysed and subsequently reabsorbed) in the hind gut, so it seems that the biliary secretion of glycopeptides represents a total loss to the animal. There is some evidence that antarctic fish have lower drinking rates than temperate fish (which would ease the antifreeze loss); this may simply reflect the generally rather higher plasma osmolarity of high latitude fish (approx. 600 mOsm. kg^{-1} in Antarctic nototheniid teleosts), which tends to reduce the rate of osmotic water loss.

Regulation of the concentration of plasma antifreezes has attracted some attention, particularly in the case of those species which exhibit seasonal changes in antifreeze concentration (e.g. Lin, 1979). Duman and DeVries (1974) demonstrated that a range of northern fish which possessed antifreeze in the winter but not in the summer, lost their antifreeze in response to a combination of heightened environmental temperature and long photoperiod; neither factor operating alone was sufficient stimulus. As Duman and DeVries point out, this sensitivity to a combination of two partially independent environmental stimuli, is effectively a fail safe mechanism, preventing an inappropriate response (for example to a situation where a lengthening spring photoperiod is associated with the coldwater aftermath of a severe winter). These workers also found that complete loss of antifreeze took 3–5 weeks.

Fletcher (1979) found that summer loss of antifreeze in the winter flounder *Pleuronectes americanus* did not occur if the fish was hypophysectomized (i.e. the pituitary gland was removed). If homogenates of pituitary were regularly injected into the peritoneal cavity of hypophysectomized fish during the summer, there was a significant decline in plasma antifreeze level within 12 days, so it would seem that pituitary hormones suppress the synthesis of antifreeze proteins by the liver during summer. Little further evidence concerning control of antifreeze concentration is available, although Lin (1979) and Lin and Long (1980) have demonstrated seasonal changes in the concentration of antifreeze messenger RNA in the winter flounder, indicating that there is some degree of control at the transcription level.

It is interesting that several species have very similar levels of antifreeze in their plasma (about 30–40 mg ml^{-1}). The mechanisms controlling concentration may be obscure, but the reason for such similar levels seems straightforward. Figure 7.5 is taken from data presented by Raymond and DeVries (1977) who measured freezing point depressions produced by different concentrations of antifreeze collected from various fish species. In each case the freezing point curves tend towards plateaux at concentrations of around 30–40 mg ml^{-1}, indicating that further increases in concentration would yield little benefit in terms of lowered plasma freezing point.

Some consideration has been given to the evolution of antifreezes, which must have occurred after the Jurassic period (approx. 300 million years ago) which appears to have been characterized by warm tropical seas (approx. 25–30°C) from pole to pole. The current antarctic climate appears to have existed for at least 25 million years, but in terms of the total duration of teleost evolution, antifreezes have appeared recently. DeVries (1976) remarks that teleosts could have responded to low temperatures by heightened plasma concentrations of ions or small organic molecules instead of the elaboration of peptide or glycopeptide macromolecules, but that the 'most conservative pathway' to freezing avoidance involved the evolution of antifreezes, because the alternative

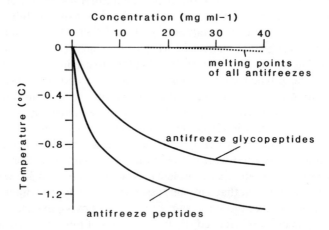

Figure 7.5 Freezing points of antifreeze solutions of various concentrations. These curves indicate that antifreeze peptides are more effective at a given concentration than antifreeze glycopeptides, but in the case of both types there is little increase in effectiveness beyond a concentration of 30–40 mg ml^{-1}. The melting points of the antifreeze solutions confirm that the antifreeze action has nothing to do with colligative properties of solutions. Redrawn from DeVries (1980).

pathways would have involved too widespread alterations in enzymes and structural proteins of the metabolic and nervous systems. This comment is fair enough, and there is certainly plenty of evidence to support the concept of low plasma ionic concentration being a conservative feature of teleost physiology (otherwise one would have expected marine teleosts in general to revert to the high blood osmolarity of their remote invertebrate ancestors, instead of conserving the low concentrations of their more immediate freshwater/brackish water teleost forebears). However, it is still necessary to identify the pre-adaptations which selection pressure could operate upon to yield the high concentrations of appropriate macromolecules. The observation that glyco-peptides and peptides of similar composition (even to the extent of featuring several fractions of different molecular weight) occur in fish species separated considerably both in geographical and phylogenetic terms, suggests that synthetic pathways of great antiquity may be involved.

From an evolutionary viewpoint it is probable that the development of antifreezes was essential to the invasion of cold, productive shallow waters by teleost fish. At one time it was generally thought that high latitude waters were universally productive (this misconception stemmed in large measure from the fact that the majority of scientists visited the polar regions in summer and mostly remained close to land). It is now known that production in large areas of the Southern Ocean is low because the phytoplankton is regularly forced down to depths below the well-lit euphotic zone (in which the breakdown of organic carbon by respiration is exceeded by the photosynthetic formation of organic carbon) as a result of violent turbulent mixing (to depths of as much as 600 m), driven by the prevailing gale-force winds. Much phytoplankton is destroyed in this manner and yearly averages of production in the Indian Ocean sector of the Southern Ocean (approx. $0.1\,\mathrm{g\,C}$ fixed $\mathrm{m}^{-2}\mathrm{d}^{-1}$) are similar to those prevailing in the nutrient-limited mid oceanic gyres of the tropics. Most of the Arctic Ocean is permanently ice-covered, so again average productivities are low. However, in sheltered inshore waters productivity can be extremely high in polar regions during spring and summer (<3.2–$3.6\,\mathrm{g\,C}$ fixed $\mathrm{m}^{-2}\mathrm{d}^{-1}$), rivalling the upwelling areas off Chile and Somalia. This primary production can support large populations of animals — provided that they can survive prevailing low temperatures. It is tempting to speculate that the near-universality of antifreezes in Antarctic teleosts (compared with a lower frequency of occurrence in the north), reflects the greater productivity and sustained subzero temperatures of the waters around Antarctica. However, the greater age of the Antarctic ecosystem may simply have allowed more time for the evolution of specialized endemic species.

7.5 EGGS OF THE CAPELIN

The capelin or lodde, *Mallotus villosus* is a small salmonoid teleost (adults are about 20 cm in length) which is of great commercial importance, millions of tons being caught each year by Iceland and Canada (and formerly by Norway and the USSR before the Barents Sea populations crashed in the mid 1980s), for the production of frozen roe, fish meal and oil. Capelin are closely related to the freshwater smelts and rather more distantly to salmon and trout. They are wholly marine, with a circumpolar distribution, but in common with other salmonoid fish probably evolved from freshwater ancestors rather recently (in evolutionary/geological terms). From a cold tolerance point of view, interest in the species largely stems from its reproductive strategy. Capelin lay large numbers of demersal eggs (i.e. eggs laid on the bottom) in early Spring. In many of the populations of capelin the eggs are laid subtidally, but in Iceland, Labrador, Greenland and Balsfjord (near Tromsø in northern Norway) populations of the fish spawn between the tidemarks, their sticky eggs being attached to gravel, stones and weed. Somewhat surprisingly the eggs are not laid when the tide is in; on rising tides males and females leap out of the advancing waves onto the substratum where copulation takes place in air, a procedure elegantly described by Jeffers (1931). Most eggs are laid on the middle shore, but substantial numbers are often found on the upper shore. They are therefore exposed for many hours each day to the influence of high latitude air temperatures which will often be well below the freezing point of sea water and fish plasma (though these temperatures will be ameliorated to some extent by the thermal inertia of the substratum on which the eggs are laid). It should be noted that the eggs are not resistant to desiccation, but survive aerial exposure because there is always a thin capillary film of water around them (Davenport and Vahl, 1983). The cold tolerance of capelin eggs was studied initially in the Balsfjord stocks, which are believed to be separate from the population of the Barents Sea (Friis-Sørensen, 1983).

Davenport *et al.* (1979) established that capelin eggs were amazingly resistant to low temperatures. They showed that the capelin egg could supercool even in the presence of ice, and in short term rapid-cooling experiments they were able to demonstrate that eggs at all stages of development did not freeze until about $-11°C$, even though ice was present in the water film around the eggs at all temperatures below $-2.5°C$. Supercooling was confirmed by recordings of egg temperature; the recording probe in contact with the egg always registered a transient rise in temperature when the egg suddenly froze and became opaque with ice. In complete contrast, newly hatched larvae froze instantly in the presence of ice (as do adult fish, whose plasma has a freezing point of -0.98 to $-1.47°C$ according to Osuga and Feeney (1978)). Further experiments showed that an intact chorion (eggshell) was needed to prevent freezing of the embryo. If a minute hole was cut in the chorion with micro-scissors, the embryo froze as soon as the environmental sea water contained ice (at around $-2°C$). On one occasion, by chance, an embryo being cooled started to hatch as the temperature fell below $0°C$ and the tip of its tail protruded through a small hole in the chorion. Ice started to appear in the (supercooled) surrounding sea water at $-3.6°C$ and the partially hatched larva immediately froze — tail first! Clearly there is no antifreeze involvement, and the chorion somehow prevented inoculation of the egg contents by ice crystals. Scanning electron microscopy studies demonstrated that the chorion of capelin eggs does indeed differ from that of other teleosts. The chorion of *Mallotus* has a sticky, thick extra outer layer, not found in other species, which appears to have the dual function of slowing ice crystal penetration (perhaps because of the small pore size of the chorion as well as the extra thickness) and attaching eggs to one another or to the substratum.

Davenport and Stene (1986) studied Balsfjord capelin egg low temperature tolerance in more detail. First they demonstrated that eggs tolerated tempera-tures at least down to $-5.2°C$ for the sort of periods likely to be encountered by eggs on the shore (Table 7.3). Second they found that there was an osmotic factor in the cold tolerance phenomenon. Although capelin eggs are extremely euryhaline, and develop normally even when regularly exposed to pure fresh

Table 7.3 Survival time of eggs of the capelin (*Mallotus villosus*) at $-5.2°C$ and $-9.2°C$ in frozen seawater films

Exposure time (h)	*Survival (%)*	*Exposure time (h)*	*Survival (%)*
At $-5.2°C$		At $-9.2°C$	
1	100.0	1	54.5
2	100.0	1.5	54.5
3	100.0	2	36.4
4	100.0	3	25.0
6	100.0	4	7.7
7	90.9	6	0.0
9	92.3		
12	90.9		
13	54.5		
14	42.9		
17	10.0		
18	60.0		
19	0.0		
22	0.0		
23	0.0		
25	0.0		

Davenport and Stene, 1986.

Table 7.4 Survival time of eggs of the capelin (*Mallotus villosus*) at $-5.2°C$ in frozen distilled water films

Exposure time (h)	Survival (%)
15 min	100.0
30 min	9.1
1 h	0.0
2 h	0.0

Eggs survive in unfrozen distilled water for as much as 4 weeks
Davenport and Stene, 1986.

water for many hours, they cannot survive freezing temperatures whilst held in fresh water films (Table 7.4). At this point it is necessary to digress somewhat to describe the general structure and osmotic features of teleost eggs. Shortly after an egg is laid it is activated (usually by the penetration of a spermatozoon) and a phenomenon known as the 'cortical reaction' occurs. The cortical (outer) layer of the egg material ('ovoplasm') secretes a colloidal material beneath the chorion. The colloidal material swells, probably by adsorption of water (Davenport *et al.* 1981), creating a space ('perivitelline space') separating the chorion from the outer membrane of the ovoplasm ('vitelline membrane'). The perivitelline space is created in some species at the expense of the ovoplasm, which shrinks, but in most species the chorion stretches. The volume of the perivitelline space in relationship to total egg volume is very variable; in cod eggs it is 15% of egg volume, while in those of the long rough dab it is huge, exceeding 80% (Lønning and Davenport, 1980). No accurate measurements have been made on the eggs of the Balsfjord capelin, but the perivitelline space of Icelandic capelin eggs is large ($70\pm6\%$ of egg volume) and the chorion very malleable (Davenport, 1989).

Once the perivitelline space has been formed, it is filled with a fluid which is close to iso-osmocity with the external medium (Holliday, 1969); this contrasts strongly with the osmolarity of the ovoplasm, which is always similar to that of the maternal blood (i.e. 300–500 mOsmoles). Generally speaking it is evident that the chorion of teleost eggs is readily permeable to salts and water, but that the vitelline membrane either has a low permeability and possesses salt pumps, or may be virtually impermeable as in the case of plaice, cod, lumpsucker and capelin eggs (Holliday and Jones, 1967; Davenport *et al.*, 1981; Kjørsvik *et al.*, 1984; Davenport and Stene, 1986).

Besides finding that capelin eggs survived low temperatures in seawater films but not freshwater films, Davenport and Stene also noted that live eggs in frozen seawater shrank dramatically, the chorion becoming crumpled and the perivitelline space becoming much smaller (although the size of ovoplasm or embryo was unaffected). In extreme cases the eggs were grossly distorted, and the chorion virtually moulded to the underlying embryo. On warming, the eggs swelled and resumed their normal spherical shape within a few minutes of all ice melting. In contrast, dead eggs were always spherical, as were all eggs (live or dead) held in freshwater films. Given these various observations, a model of freezing resistance was produced (Figure 7.6). It seems that when ice starts to form in seawater films (at about −1.9°C under field conditions), the salinity (and therefore osmolarity) of the unfrozen water rises. Given the

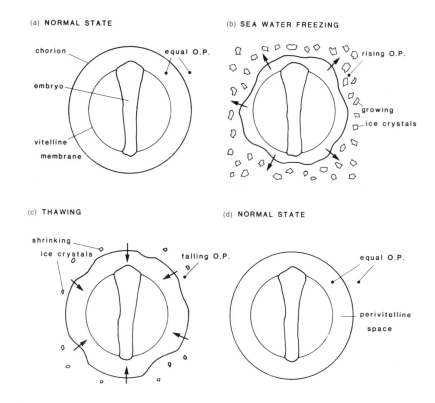

Figure 7.6 Model of freezing resistance in eggs of the capelin (*Mallotus villosus*). In (a) the egg is in sea water and the perivitelline fluid between the chorion and vitelline membrane has the same osmotic pressure (OP) as the sea water. In (b) the sea water has begun to freeze. The salinity of the unfrozen sea water rises and water moves osmotically across the highly permeable chorion, causing the egg to shrink, the egg exhibiting large surface depressions. In (c) the frozen sea water thaws and the salinity of the unfrozen sea water falls. In consequence water flows back into the perivitelline space and the egg resumes its normal shape (d). The model requires that the vitelline membrane be impermeable to salts and water, that the chorion be flexible and permeable to both salts and water (but more permeable to the latter). Available evidence supports all of these requirements. From Davenport (1989).

known high permeability of the chorion to water (Davenport and Vahl, 1983), water will flow rapidly out of the perivitelline space into the hyperosmotic outside medium. This osmotic water flow will have two consequences. Firstly the egg will shrink; secondly the osmolarity of the perivitelline fluid will rise and therefore the fluid will become less likely to freeze. The high osmolarity perivitelline fluid will effectively form a barrier between the supercooled embryo and the outside world. In contrast, a capelin egg cooled in a distilled water film will have no tendency to lose water when freezing starts. Indeed, the perivitelline space will tend to take up water and lose salts, so that the perivitelline fluid becomes more likely to freeze.

This model demands that the capelin egg should have a tough and elastic chorion which does not readily allow the passage of 'seeding' ice crystals; it also demands a large perivitelline space. Both chorion elasticity and a large perivitelline space are features which are characteristic of salmonoid eggs which are generally laid in sand/gravel subject to strong water movements. These features provide protection against damage by abrasion and movements of stones and pieces of gravel, but are also preadaptations for cold tolerance. Davenport *et al.* (1986) investigated the mechanical characteristics of the chorions of a variety of teleost eggs and found that the capelin chorion was unusually flexible, yet very resistant to tearing.

The model of Davenport and Stene required that salt fluxes across the chorion should be slower than water movements; no data were available concerning this point, but it was difficult to see how egg shrinkage and swelling could occur if the chorion were not at least partially semipermeable. Finally, it would be necessary for the vitelline membrane to have a low permeability to salts and water, or the embryo would become osmotically dehydrated during the shrinking process. The euryhalinity demonstrated by Davenport and Stene supported this concept of low permeability.

More recently still, Davenport (1989) has shown that eggs of Icelandic capelin (spawning subtidally) are just as tolerant of low temperatures (6 h median lethal temperature = $-6.3°C$) as the Balsfjord eggs, so it seems unlikely that the Balsfjord fish represent a physiological race. Davenport (1989) demonstrated that capelin eggs show great tolerance of high salinities. Eggs exposed to high salinities for 6 hours showed a median lethal salinity of 88.3‰. He also showed that eggs transferred from sea water to 72‰ first shrank dramatically (showing huge surface depressions in 10 s — (Figure 7.6), but then resumed their spherical shape over about 260 s, during which time the embryo gradually floated to the top of the egg as the concentration of the perivitelline fluid rose. This indicated that the chorion was extremely permeable to water (hence the rapid shrinkage), but rather less permeable to salts — precisely the features demanded by the model of Davenport and Stene. Davenport (1989) also showed that embryos frozen in freshwater films died because their tissues were disrupted by ice crystals. In contrast, those embryos which finally succumbed at low temperatures in seawater films died because of osmotic dehydration — presumably because the concentration of the perivitelline fluid eventually rose to intolerable levels.

7.6 FREEZING IN INTERTIDAL INVERTEBRATES

The intertidal zone of temperate latitudes is occupied by a host of sedentary and sessile animals which cannot migrate into deeper water to avoid exposure to low temperature when the tide ebbs. Barnacles, bivalve molluscs and gastropods form the bulk of the animals which have attracted study. It can be difficult to be sure of the temperatures that such animals are likely to encounter in cold weather. Ice and snow often blanket organisms and protect against low air temperatures — a sort of greenhouse phenomenon (Williams, 1970), while the substratum (whether rock, sand or mud) is likely to take much time to cool, so the animals living on or in it may be exposed to a narrower range of temperatures than the atmosphere. Substrata high in the intertidal zone will reach more extreme temperatures than substrata near the low water mark. Masses of macroalgae may function in fur-like fashion to keep small animals (e.g. littorinid gastropods) relatively warm for some time after the tide ebbs. Heat is likely to be stripped from coarse-grained substrata (which drain readily) more quickly than from substrata which hold water and therefore have a great heat capacity. It is consequently clear that a mosaic of microclimates exists on the shore, so that members of a population of invertebrates will not all encounter the same extremes of temperature under adverse weather conditions. Unusually cold weather undoubtedly kills substantial numbers of intertidal invertebrates. The severe UK winters of 1947 and 1963 caused massive losses in many populations (e.g. Crisp *et al.*, 1964), but it is well known that a number of intertidal animals survive prolonged exposures to temperatures of $-10°C$ or lower. These temperatures are far below the likely freezing point of invertebrate body fluids (approx. $-1.9°C$). Supercooling is not likely to be an option available to many such invertebrates because their external body surfaces and gut linings are coated in silt and food material likely to promote freezing by inoculation (Chapter 6). Observations long ago established that intertidal invertebrates tolerated freezing of the body fluids, and survived provided that freezing was limited to the extracellular compartment (e.g. Kanwisher, 1955). Except under very unnatural laboratory conditions (cooling at 100–700 deg. C. min^{-1}, yielding minute crystals which do not disrupt subcellular organelles — Mazur, 1984), no cells have been found to tolerate ice in the intracellular compartment. Later work also demonstrated that freezing injury was not normally due to intracellular freezing, but to mechanical damage by external ice crystals, and to osmotic dehydration as the concentration of unfrozen extracellular fluid rose, thus creating an osmotic gradient across the cell membranes. Mortality has been variously attributed to lethally high salt concentrations (Lovelock, 1953), protein denaturation resulting from the removal of so called 'bound water' (Karow and Webb, 1965) and reduction in cell volume to a critical level (Meryman, 1971). These hypotheses are not mutually exclusive; all antedate modern knowledge of the structure of cell membranes, and it seems probable that an additional cause of mortality may stem from compression of the cell membrane (as its surface area is reduced), eventually resulting in a lethal change in its ordered nature. Storey and Storey (1988) point out that extracellular ice formation is damaging for two reasons; first because small ice crystals coalesce to form larger, more disruptive crystals ('ripening' or 'recrystallization' — Chapter 1), second because the extracellular ice crystals isolate cells and inhibit blood flow, depriving them of oxygen and nutrients. They believe that adaptations to cope with these aspects of extracellular freezing are of crucial importance to survival of intertidal invertebrates at subzero temperatures.

Surprisingly low temperatures are tolerated by such animals; the periwinkle *Littorina littorea* has been found to survive to $-22°C$ with 76% of its body water frozen (Kanwisher, 1955). Crisp *et al.* (1977) reported that the boreal barnacle *Semibalanus balanoides* survived temperatures as low as -18 to $-20°C$ (the precise value varying from year to year) in winter; they found that barnacles sampled in January 1973 had a median lethal lower temperature of $-18.3°C$ at which temperature more than 80% of the body water was frozen.

Work over the past 20 years has shown that the precise lower lethal temperature of a given invertebrate depends on many factors. Williams (1970) developed the concept that mortality occurred when a certain proportion of the body water (about two thirds) was frozen. This over-simple idea has been denied by subsequent work. Murphy (1979) showed that there were differences in lower lethal temperature of gastropod species related to their heights on the shore. Littorinids from high on the shore tolerated much lower temperatures than did whelks from the low water spring tide level; they also tolerated a much higher proportion of their body water being frozen (82% vs 65%). Several studies have shown that there are usually marked seasonal differences in lower lethal temperature. An extreme example is the barnacle *Semibalanus balanoides* which dies in summer at $-6.6°C$ with about 40–45% of body water frozen and at $-18.3°C$ in winter with over 80% frozen (Crisp *et al.*, 1977). The mussel *Mytilus edulis* has yielded much useful information in this area. Theede (1972) demonstrated that lower lethal temperatures of mussels depended on season, acclimation temperature and salinity of habitat (with animals from full sea water tolerating lower temperatures than animals from the brackish Baltic). He also showed that that complete acclimation between maximum and minimum tolerance of freezing temperatures was accomplished in some four weeks.

Of particular interest is the identification of the factors which separate freezing-tolerant and freezing-intolerant invertebrates. Storey and Storey (1988) have recently written a review which addresses this problem. They believe that the crucial difference lies in the elaboration of two types of protein by freezing-tolerant animals (both intertidal and terrestrial). The first type, ice-nucleating proteins (INPs), have the function of initiating freezing in the extracellular body fluids. Such nucleating proteins are common in insects and are discussed in more detail in Chapter 6. However, they are important to intertidal animals because they allow freezing to be initiated as soon as possible after the body temperature has fallen below about $-1.9°C$. This allows freezing to proceed slowly, preventing the build up of damaging osmotic gradients across cell membranes. The second type of proteins are thermal hysteresis proteins (THPs), again also found in terrestrial arthropods (Chapter 6), but in that case associated with avoidance of freezing. THPs were identified several years ago in *Mytilus edulis* (Theede *et al.*, 1976) and their function at first seemed anomalous — why would a freezing-tolerant animal need antifreeze proteins? However, it is now apparent that THPs have an additional function; they prevent ripening or recrystallizing of ice crystals. They therefore inhibit the formation of dangerously large ice crystals which would disrupt tissues. THPs have also been found in the Antarctic limpet *Nacella concinna*, where they were originally reported to prevent freezing in conjunction with copious secretion of mucus containing THPs (Hargens and Shabica, 1973). As Antarctic limpets are freeze-tolerant (Davenport, unpublished data), it is probable that they also have the function suggested by Storey and Storey.

Storey and Storey (1988) list four other requirements for freezing tolerance, none of which are unique to freezing-tolerant animals. The first of these is that cell volume must be regulated. Freezing-tolerant invertebrates inevitably suffer some cell shrinkage, but this is controlled to some extent by the free amino acid pool within the cell (which promotes a high intracellular concentration of non-permeant, osmotically active molecules, thereby restricting water flow out of the cell). In mussels, Williams (1970) demonstrated that acclimation to high salinities enhances survival at subzero temperatures. Acclimation to high salinities over a period of a couple of days causes an increase in size of the free amino acid pool, which therefore acts in a cryoprotective manner. The 48 hour period of acclimation to salinity is essential, since shorter periods of exposure to high salinity themselves result in cell shrinkage, not reversed until the size of the intracellular amino acid pool is raised. Mussels and cockles (*Cerastoderma edule*) acclimated to low salinities have a much reduced toleration of freezing (Figure 7.7), presumably because of a reduced level of cryprotective amino acids within the cells (Ibing and Theede, 1975). Regulation of intracellular amino acids to control cell volume is widespread in euryhaline invertebrates exposed to wide salinity ranges (see Rankin and Davenport, 1981 for review), and all of the freezing-tolerant

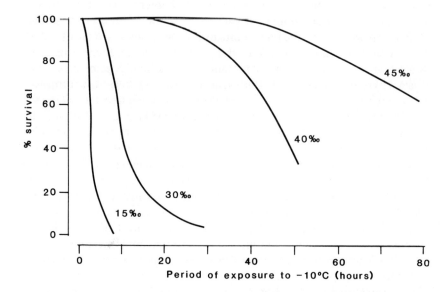

Figure 7.7 Effect of acclimation to various salinities on freezing resistance in the cockle, *Cerastoderma edule*. The data show that acclimation to high salinity (for a long enough period to permit changes in the concentration of intracellular free amino acids) substantially improves freezing resistance. Redrawn from Theede (1972).

intertidal invertebrates appear to be euryhaline; clearly a euryhaline physiology is a good preadaptation for low temperature survival.

Next, a freezing-tolerant animal must maintain membrane function even at substantial subzero temperatures ('membrane stabilization'). Physical functions such as gaseous and ion diffusion, transport functions such as ion pumping, plus receptor and neural functions must proceed virtually independently of temperature. Maintenance of fluidity of membranes is a factor in this stabilization, and seasonal restructuring of membranes ('homeoviscous adaptation'), perhaps by incorporation of more polyunsaturated fatty acids into phospholipid membranes in winter, has been suggested as a likely adaptation, but available evidence is as yet inconvincing. Franks (1985) indicates that homeoviscous adaptation is important to maintenance of membrane fluidity at low temperatures, but that this fluidity is not related to freezing resistance *per se*. Storey and Storey (1988) suggest that membranes must be adapted to prevent transmission of ice crystals from the extracellular compartment to the cell contents; perhaps THPs have a rôle in this respect.

The third requirement of a freezing-tolerant animal is that it must have an effective anaerobic metabolism. The formation of ice crystals in the extracellular fluids causes anoxia and ischaemia, so the cells must be able to function in a low oxygen tension environment. Again, sessile intertidal invertebrates appear to be preadapted to deal with this situation. Most have behavioural responses which isolate them from the environment under adverse conditions. For example: the mussel *Mytilus edulis* isolates itself by shell valve closure in response to emersion, low salinity and some pollutants (Davenport, 1977, 1979a, 1981). Oxygen tensions within the mantle cavity fall with a few minutes to negligible levels (Davenport, 1979) and anaerobic by-products of metabolism are identifiable within a few hours (de Zwann, 1977). Anaerobic glycolysis in mussels is particularly efficient, creating about twice as many ATP molecules (or equivalents) from a given amount of glucose as mammalian muscle glycolysis. Barnacles, gastropods and bivalves such as cockles, clams and oysters show similar isolation behaviour.

Finally, freezing-tolerant invertebrates need to switch to a reduced metabolic rate ('torpor') to eke out fuel which cannot be readily replaced during the period of freezing. Contrary to popular opinion, metabolism is not abolished at 'moderate' subzero temperatures (-2 to $-30°C$); only at temperatures of $-196°C$ (the temperature of liquid nitrogen) does this happen. However, freezing-tolerant animals all demonstrate switching to low metabolic rate processes, usually before they start to freeze. Kanwisher (1959) found that the periwinkle *Littorina littorea* entered a dormant state at $0°C$, when the Q_{10} of oxygen uptake rose from the normal 2–3 to nearer 50. Murphy (1983) found that the mussel *Modiolus demissus* entered a dormant state when it was cooled below $4°C$, and that the species had a Q_{10} of 22 between 0 and $4°C$. Entering a dormant state may reduce cellular energy demands by 90%. Murphy demonstrated that allowing *Modiolus demissus* to enter a hypometabolic state slowly (by holding at $<4°C$) improved subsequent survival at freezing temperatures.

7.7 FREEZING AVOIDANCE BY INTERTIDAL INVERTEBRATES

In the late winter of 1979 the author discovered a strange transient environment on the subarctic shores of the Tromsøsundet in northern Norway, near the marine biological station of the University of Tromsø. Shallow intertidal pools on the middle and upper shores froze over on each low tide, since air temperatures were below $-10°C$. The pools had a simple fauna, consisting of three species, the gammarid amphipod *Gammarus duebeni*, which occurred in high numbers, a flatworm (*Procerodes ulvae*) and small specimens (1–2 cm shell

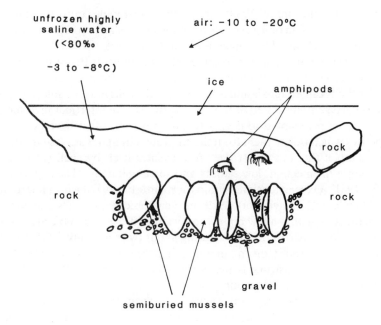

unfrozen highly
saline water
(<80‰

−3 to −8°C)

air: −10 to −20°C

ice

amphipods

rock

rock

rock

gravel

semiburied mussels

Figure 7.8 High latitude rock pool during low tide in winter (data collected by the author in subarctic Norway). See text for details.

length) of the mussel *Mytilus edulis*. All three species occurred in great densities, and it seems probable that the gammarids used the pools as refuges as the tide ebbed, since there was little material in the pools for them to eat. The situation of the small mussels was unusual. Most mussels are epifaunal, but the small mussels in these tidepools were buried, tightly packed, in coarse gravel at the bottom of the pools. Only the posterior margin of the shells protruded above the surface of the substratum. As the pools froze, a hypersaline environment developed beneath the ice (Figure 7.8). Davenport (1979b) reported salinities of 50–60‰ when the ice was about 1 cm thick, but on subsequent occasions (e.g. Davenport and Carrion-Cotrina, 1981), salinities of more than 80‰ have been recorded. It seems likely that there is much small-scale variation in salinity beneath the ice, as Schlieren lines could be seen in the wakes of swimming gammarids. Recorded temperatures in the hypersaline water of these intertidal pools ranged from −3 to about −8°C. Interest initially focussed on the amphipods, which were often seen to swim slowly beneath the ice of their pools (Davenport, 1979b), though the bulk of the population aggregate beneath stones at the bottom. To study the survival mechanism, the response of amphipods to two simulated pools (Figure 7.9) was investigated. In a 'normal' pool, where ice formed at the surface, the gammarids aggregated

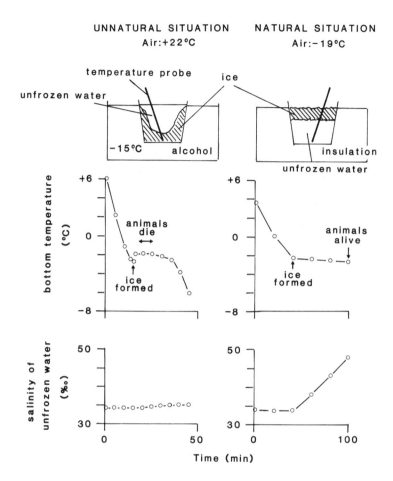

Figure 7.9 Low temperature survival in the high latitude intertidal amphipod *Gammarus duebeni*. Small simulated pools containing sea water and amphipods were cooled. On the right the pools were cooled in a natural manner so that ice formed on the water surface. The salinity of the unfrozen water beneath the ice rose steadily. The amphipods aggregated at the bottom of the pool and survived. On the left an unnatural situation was contrived. The 'pool' was held in a cold alcohol bath in a warm room; when the temperature of the sea water dropped to −1.9°C ice started to form at the bottom of the pool. The amphipods swam downwards and froze after some seconds' contact with the ice; they did not recover. From Davenport (1979b).

at the bottom in water of rapidly increasing salinity. Occasionally they would swim up to the lower surface of the thickening ice, but did not freeze, even though its temperature was about −4.7°C. In contrast, when a situation (impossible in nature) was produced in which the 'pool' froze from the bottom up, and the unfrozen sea water showed little salinity increase, the amphipods froze and died. Davenport also found that specimens of *Gammarus duebeni* would only freeze in air at −10°C if they were pressed against ice for a few seconds. In addition, he observed that amphipods coming into contact with large ice crystals would kick themselves away. In summary, it appeared that the gammarids survived in the frozen rockpool environment, partly because migration of ice crystals through the integument took an appreciable amount of time, and also because most of the animals stayed in an unfrozen blanket of hypersaline water at the bottom of the pool, thus avoiding sustained exposure to ice. Survival in this high salinity environment demands euryhalinity. Davenport (1979b) measured blood concentrations of amphipods exposed to a wide range of salinities for various periods of time (Figure 7.10). These data confirmed that *Gammarus duebeni* was a powerful osmoregulator which showed relatively little change in blood osmolarity, especially after short-term exposure. The results also demonstrated that *Gammarus duebeni* acclimated to full sea water were slightly hyperosmotic to that medium (by about 70 mOsm kg^{-1}). This means that the amphipods' blood is not immediately vulnerable to

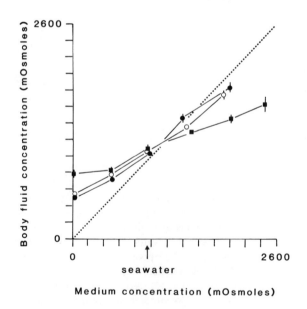

Figure 7.10 Osmotic physiology of *Gammarus duebeni*. Amphipods were placed in water of a range of salinities and their body fluids sampled for osmolarity after 2 hours (squares), 24 hours (open circles) and 7 days (closed circles). The dashed line represents equality of osmolarity. The data shows that the species is a powerful hyper-hypo-osmotic regulator which is hyperosmotic to sea water, so not susceptible to freezing. From Davenport (1979b).

freezing as soon as the pools start to freeze — allowing time for them to respond to falling temperatures by aggregation at the bottom of the pool.

Davenport and Carrion-Cotrina (1981) investigated the situation of the mussels living in the gravel bottom of the pools. Again a laboratory simulation approach was adopted, a simulated tide pool being manufactured which allowed the repeated exposure of immersed mussels to a combination of rising salinity and falling temperature (Figure 7.11). Experience of real and simulated pools showed that mussels gained considerable insulative advantage from living in rock pools. By living at the bottom of pools mussels avoided temperatures lower than about −6 to −8°C, and thus survived periods when air temperatures fell as low as −20°C — well below the lower lethal temperature of *Mytilus edulis* (roughly −10 to −15°C). Further study showed that the mussels isolated themselves from the environment when water temperatures dropped to 0°C. At this temperature the siphons were closed and the animals stopped pumping sea water. The mussels adducted their shell valves tightly at −1.5°C. Throughout the period when tidepool salinities were rising quickly, the mussels (which are osmoconformers) retained normal sea water around their tissues, thereby avoiding osmotic stress; they also remained unfrozen and were supercooled. Clear evidence that the mussels closed in response to falling

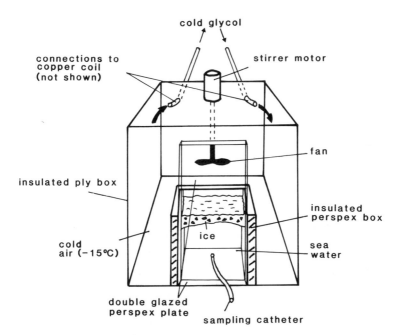

Figure 7.11 Design of artificial high latitude rock pool used in studies upon the mussel *Mytilus edulis* (Davenport and Carrion-Cotrina, 1981).

temperature rather than rising salinity was presented. During thawing, foot movements and pumping were restarted at $-1.5°C$, indicating that this behaviour pattern maximizes access to food and oxygen.

Davenport and Carrion-Cotrina also investigated heart rate and activity of the frontal gill cilia. When Norwegian mussels were cooled from $+18°C$ to $-1.5°C$, the heart rate steadily fell from 30 beats min^{-1} to 3 beats min^{-1} ($Q_{10} = 3.3$). When the shell valves were shut at $-1.5°C$ the heart beat became irregular and stopped altogether after a few minutes. Ciliary activity also fell to negligible levels at temperatures below $0°C$, so it is apparent that mussels are in a torpid state when the tide pools are frozen over.

PART FOUR

Man and Cold

8 Man and cold

8.1 INTRODUCTION

There is no doubt that *Homo sapiens* is physiologically a tropical species, as originally pointed out by Per Scholander (he applied the adjective to Inuit!). The remarkable series of palaeontological discoveries by the Leakeys in Kenya and Tanzania tend to pinpoint the origin of man to tropical Africa, but the lower critical temperature of humans (LCT = 27°C resting in air; 33°C resting in water) is proof positive, since it is similar to that of tropical animals such as the two toed sloth (*Choloepus hoffmanni*) and night hawk (*Nyctidromus albicollis*) (Scholander *et al.*, 1950a–c). In contrast, arctic birds and mammals have LCT values at least as low as 10°C, and in some species (such as the arctic fox *Alopex lagopus*) the thermoneutral zone may stretch to temperatures well below 0°C (Korhonen *et al.*, 1985). A naked man shows increasing heat production as environmental temperatures fall (Figure 8.1), needing twice as much energy to maintain a constant core temperature of 37°C at 10°C as at 20°C. Heat production may be by muscular thermogenesis (mainly shivering) or by non-muscular/non-shivering thermogenesis (NST), principally by the liver.

Another feature which identifies Man as a tropical animal is his relative hairlessness. There has been much argument concerning the selective pressures involved in the evolution of short hair in humans. Currently the generally accepted hypothesis is that it aids heat loss during exercise in hot climates, and probably arose as hominids moved from wooded country to plains. However, the lack of an insulating furry coat undoubtedly contributes to Man's inability to tolerate low temperatures.

Currently, people live at virtually all latitudes (including those of Antarctica) and most altitudes, living in environments as cold as −80°C in winter. Man can even live for long periods in the cold of space (approaching absolute zero, −273°C). While some of this increased area of distribution is undoubtedly due to modern technology, it is salutary to remember that the Inuit (formerly

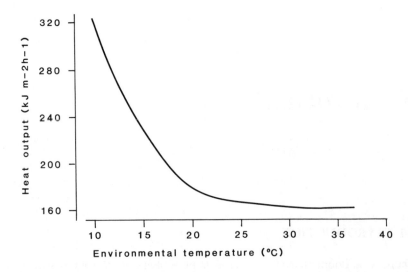

Figure 8.1 Change in heat output by naked man exposed to falling environmental temperatures.

known as Eskimos) and Saami (often referred to inaccurately and offensively as Lapps) of northern Europe, Asia and America have made their living beyond 70° North for at least 6000–10 000 years. Earlier races were hunting (and possibly taming) reindeer in northern Europe 25 000 years ago, while *Homo sapiens neanderthalis* endured the cold of the Ice Age 100 000 years ago, along with his thick-furred pachyderm prey. Australia aborigines and Kalahari !Kung tribesmen have coped with the near-freezing cold of desert nights for millenia. These various peoples survived and survive by possession of one or more of the following: fire, clothes or weather-proof habitation.

8.2 HUMAN MORPHOLOGY AND COLD

Loss of heat from the body of an endotherm exposed to low temperatures is dependent to some extent on its surface area to volume ratio. *A priori* one would expect selection in high latitude mammals and birds to minimize this ratio. The utility of Bergmann's rule (for body size) and Allen's rule (for dimensions of extremities) reflect such selection. However, as already discussed in Chapter 4 it is simplistic to consider endotherms as simple, warm bodies; they have considerable physiological control over heat loss (particularly by control of the peripheral circulation). Do these rules apply to *Homo sapiens*? This is a difficult area of study for a number of reasons. First, human populations have been extremely mobile historically (especially following the transport developments of the past century), so the environmental conditions which may have given rise to a particular body form or somatotype, are often difficult to discern. Second, study demands access to very large data sets. These are now difficult to obtain for financial, political or cultural reasons, so much reference has to be made to old data, often collected to support spurious racist theories. Third, there is much evidence to show that human morphology is plastic and greatly influenced by nutritional standards (in terms of quantity and quality of food). Japanese stature has changed greatly since the Second World War, and in both Europe and the USA there has been a trend towards greater adult height during the last few decades. Roberts (1973, 1978), who has worked within these constraints, demonstrated that humans tend to be heavier and to have a greater chest girth in areas where annual temperatures are low (Figure 8.2). This suggests that Bergmann's rule operates for humans. Roberts also showed that people inhabiting cold regions tended to have decreased relative limb length (and increased relative sitting height), indicating that Allen's rule is applicable too. In both cases it is evident that the data gathered are variable and

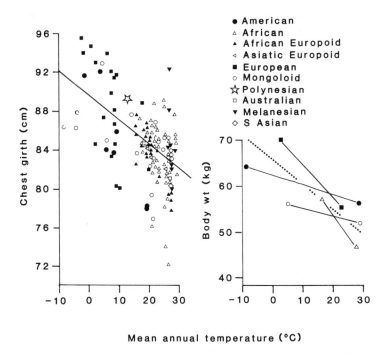

Figure 8.2 Evidence for relevance of Bergmann's Rule to human populations. These data indicate (despite variability) that chest girth and body weight increase with declining mean environmental temperature. Redrawn from Roberts, 1973.

that such trends only become obvious when large numbers of people are measured. Earlier attempts at correlating height with thermal environment were less successful — basically because tall, slender people (e.g. Nilotic Africans) may have a higher surface area to volume ratio than short, stocky people of similar body weight. Although there are indications that body weight fluctuates seasonally in humans (peaking in winter), the amplitude of fluctuation is low (<2 kg) and it seems likely that any changes in subcutaneous fat thickness are of negligible insulative effect, and that putting on weight for the winter is probably an energy-storing device. However, there is plenty of evidence that cold tolerance is improved in humans possessing significant quantities of body fat. Kollias *et al.* (1974) compared the responses of 7 obese (fat making up 29–41% of body weight) and 3 lean (fat = 21–24% body weight) women to placement in a 20°C water bath. The lean women lost heat faster and increased heat production more than did the obese women. Wyndham and Loots (1969) compared the responses of thin and fat men in cold climates, finding that the thin lost heat faster and had higher skin temperatures; they also showed more discomfort and tolerated low skin temperatures less well. Interestingly, these workers found that thin men living in the Antarctic for long periods slowly increased in weight and showed enhanced skin fold thickness (indicating fat deposition), whereas fat men showed no change.

Attempts have also been made to correlate head shape with cold adaptation. The peripheral tissues of the head are generally held at high temperatures; the blood flow to scalp and cheeks is not reduced much in response to cold. Coon *et al.* (1950) postulated that the mongoloid face, apparently originally characteristic of people in cold regions of N.E. Asia (who later spread to both north and south America as well as southern Asia), possessed features which had evolved in response to cold – flat appearance, fat padding around the eyes and nose (the latter helping to warm inspired air), plus reduced protuberances (particularly brow ridges).

Skin colour has attracted much attention in humans, dark skin being associated with tropical climates (Gloger's rule). Most workers have tended to assume that the generally light skin of high latitude humans is non-adaptive — simply not selected against, so gradually increasing in the population by genetic drift. However, it has become clear that genes for light skin are dominant and it has been suggested that light skin may have value in cold climates. The lack of melanin in the skin of whites allows the penetration of ultraviolet light (in the wavelength range 2500–3130 Å) far enough into the tissues to rearrange atoms in sterol molecules which consequently acquire Vitamin D properties and provide protection against the skeletal abnormalities of rickets. At higher latitudes, light levels are low and dietary sources of Vitamin D meagre, so this source of vitamin would appear to be of selective advantage. However, in some respects this hypothesis appears dubious. At very high latitudes humans necessarily wear all-embracing clothes through most of the year, have strongly tanned hands and faces (thus reducing ultra-violet penetration), and in any case eat diets rich in fish oils (which contain plenty of Vitamin D). It is difficult to see how light skin could be of great value to these people. Much of the evidence in favour of the hypothesis stems from the high incidence of rickets in some of the industrial cities of Europe in the early part of the present century (the industrialization being associated with dark, narrow streets and smoky atmospheres), together with the observation of rickets in present day northern cities (e.g. Glasgow) amongst immigrant popu-lations with dark skins (who presumably have difficulty in synthesizing much cutaneous Vitamin D). For both groups it is difficult to avoid the conclusion that economic and dietary deprivation contribute to the incidence of rickets, so the importance of lack of skin pigment in temperate and polar latitudes is still unclear.

8.3 PHYSIOLOGICAL/METABOLIC ADAPTATIONS AND RESPONSES

It is common for a widely-distributed animal species to exhibit clines of physio-logical response to environmental stresses. We might therefore expect to find that humans from cold climates would show physiological or metabolic adaptation to cold. In fact, it has been rather difficult to obtain clear evidence to support this hypothesis. Northern races such as the Inuit and Saami now have access to modern clothing and housing. However, even when they lived in igloos and turf houses a century ago and wore furs rather than quilted man-made fibre, they essentially survived by avoiding cold. Winters and spells of poor weather were spent huddled around fires, subsisting on hoarded food. Some evidence of heightened basal metabolic rate (non-muscular thermo-genesis) at low temperatures is evident from these groups, but it is not entirely convincing. Roberts (1978), using data from a number of studies of a wider range of indigenous humans, claimed a significant negative correlation be-tween 'basal metabolic rate' and mean annual environmental temperature, indicating that humans native to colder climates had evolved significantly higher metabolic rates. Unfortunately, Roberts defined basal metabolic rate as the rate measured at rest (not asleep) under post absorptive conditions, but at an ambient temperature of 20°C. He did not state whether the individuals studied were clothed. The ambient temperature used is, of course, appreciably below the temperature of thermoneutrality in humans (lower limit 27°C), so in fact the conditions employed could not guarantee truly basal metabolism. If the subjects were unclothed, then some of the metabolic rate would be due to regulatory non-shivering thermogenesis which is known to be more pro-nounced in cold-acclimatized animals, so one would expect higher metabolic rates from humans acclimated to cold climates — but this would not indicate underlying genetic differences. If the subjects were clothed, then interpretation becomes impossible.

Two groups of humans have been studied under conditions unaffected by post-industrial changes; both have been exposed to cold for hundreds of generations whilst largely isolated from other humans; conditions which would favour the selection of advantageous physiological characteristics.

Native Australians (aborigines) have inhabited the arid interior of Australia for at least 40 000 years. Although the desert is very hot during the day, the cloudless nights guarantee low temperatures, often near 0°C. Physiological study of aborigines and whites sleeping without clothes between small fires at night, showed that the former slept soundly without shivering, while the latter could sleep only for periods of a few minutes, interspersed with violent bouts of shivering. Measurements of rectal (i.e. core) temperatures showed that the aborigines tolerated a decline in body temperature during the night, while the temperature of the whites remains steady, but this steadiness was paid for by an increased heat production by muscular thermogenesis. Aborigines also allowed their extremities to become cold without raising the metabolism (Scholander *et al.*, 1958). There was no indication that the aborigines (or the whites) produced extra heat by non-shivering thermogenesis. Edholm (1978) suggested that the native Australians could tolerate a nighttime decline in body temperature, because this would be rapidly reversed by the morning sun.

A second group of humans exposed to cold conditions with minimal clothing and habitation cannot rely on the sun in this fashion. Tierra del Fuego ('Land of Fire') is at the tip of South America, surrounded by the Southern Ocean. Despite its name, the region is characterized by perennial cold and wet conditions, often freezing in the winter and always cloudy or misty, with frequent gales. Charles Darwin remarked on the hardiness of the people who lived there, an Indian tribe, the Alacalufe, since almost exterminated by introduced diseases and the cruelty of settlers from further north. Hammel (1964) studied surviving members of the Alacalufe and found that they did not appear to shiver during sleep, but maintained a high metabolic rate and steady body temperature throughout the night, presumably by non-muscular heat production (although in fact it is virtually impossible to guarantee absence of microshivering without electrical recordings from muscles). Edholm (1978) speculated that Alacalufe Indians could not 'afford' to cool down at night, as they would not be able to rewarm easily in the morning. He suggested that heightened metabolic rate had evolved to prevent such cooling, but a simpler explanation may be that the Alacalufe were the only humans to move into a permanently cold, wet and windy enivironment without employing adequate technological solutions; heightened metabolic rate would have been selected simply to survive, whether during the day or at night. Whatever the details of selection pressure, the physiology of the Alacalufe makes an interesting contrast to the physiology of the Australian aborigines and whites (Figure 8.3),

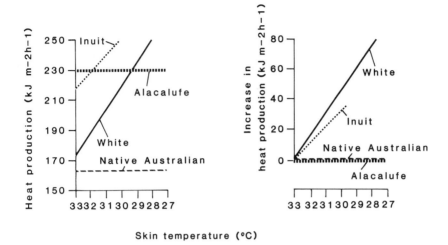

Skin temperature (°C)

Figure 8.3 Influence of declining temperature on metabolic rate in naked humans of various origins. There is relatively little difference between Inuit and Whites; both groups show heightened heat production in response to cold. Alacalufe and Native Australians do not show such increases, but have different basic strategies. Alacalufe have a high basic metabolic rate which extends their thermoneutral zone to lower temperatures than other human groups; Native Australians allow their core body temperatures to fall rather than increase in metabolic rate. Redrawn from Edholm (1978).

indicating multiple solutions to the problems imposed by cold. Roberts (1973) reinforced this view, reporting that Mongoloid peoples achieved cold tolerance by non-muscular thermogenesis, Australian aborigines and Samic people by exchange of heat between outgoing arterial blood and incoming venous blood ('countercurrent exchange'), while whites relied on the insulation of subcutaneous fat.

Much is now known about short term physiological responses to cold in humans, and there is also some information about cold acclimatization. When humans are exposed to temperatures below those of thermoneutrality, they exhibit a suite of responses (generally under hypothalamic control — Chapter 4). They become more active (thus generating heat by muscular thermogenesis) and may also exhibit involuntary movements plus shivering. Peripheral vasoconstriction occurs so that the outer 'shell' of the body becomes cooler than the core (thereby making the thermal gradient between the body and the environment much shallower), with the hands and feet becoming particularly cold. At intervals the vasoconstriction of the blood supply to fingers, toes, elbows, knees and parts of the face (including the nose) is replaced by brief vasodilatation. Vasodilatation is accomplished by reduced sympathetic output to nerves which control the muscular walls of vessels which connect small arteries and veins in these regions ('arterio-venous anastomoses'). The increased blood flow produced by the opening of these shunts rapidly warms the nearby tissues, protecting them against frostbite. Vasoconstriction and vasodilatation alternate in cyclical fashion, the net effect being to reduce heat output from the extremities without tissue damage. If cooling continues, the subject will instinctively adopt a huddled position in which the arms are wrapped around the body, the knees brought up and the head bent. This posture reduces surface area, directs the warm exhaled air towards the trunk, and keeps the groin and armpits (regions where essential blood flow is quite close to the surface) protected. These various postural changes can reduce heat loss by convection and conduction by 50% (Edholm, 1978). In males, blood flow to the penis is reduced and the cremaster muscles contract to bring the testicles closer to the abdomen. Given further cooling, a human is likely to suffer the impaired functioning associated with hypothermia as the core body temperature starts to fall. Cooling of the body also causes increased urine output and increase frequency of micturition (passing of urine). This is because production of antidiuretic hormone (ADH) by the posterior lobe of the pituitary gland is inhibited. The adaptive significance (if any) of this 'cold diuresis' is unknown — though exploited by doctors and nurses who place patients on cold floors to facilitate the production of urine samples!

Evidence of true acclimatization to cold is fairly limited. We have already seen that physiological and anatomical adaptations of Inuit, Alacalufe and native Australians to cold are relatively minor, and it has been even more difficult to discern differences in metabolism or physiology arising from exposure of humans from lower latitudes to weeks or months of cold. Slight seasonal variations in body fat content are seen in people living in temperate regions, but these are only of the order of 2 kg between summer and winter — insufficient to have significant effects on insulation. In any case the changes in body weight could easily be explained by the general tendency of people to take less exercise in winter. Experiments have also been carried out upon personnel working at Arctic and Antarctic military/scientific stations, but in fact these people take a great deal of care over the warmth of housing and accommodation, so that they are rarely exposed to low temperature, and then only briefly. Some studies have suggested that such workers at high latitude become more capable of maintaining core temperature in the face of cold, but the data are rather unconvincing. It is probable that fishermen working in the north Atlantic, shepherds working flocks in upland areas of Europe and similar people who are regularly forced out of doors in temperate winters may provide more fruitful targets for study.

There is convincing evidence of acclimatization in one sense; the progressive toleration of cold extremities for longer periods. This local acclimatization was first studied in detail by Mackworth in the late 1940s (reported in Edholm,

1978). Mackworth demonstrated that tactile sensitivity in the fingers was maintained by outdoor workers at Fort Churchill, Hudson Bay (in winter) for much longer periods under cold conditions than it was in office workers. He also demonstrated that such cold acclimatization was completed in about six weeks. The basis of this phenomenon is not entirely clear. To some extent it is due to a reduction in the depth of peripheral vasoconstriction, and a more rapid onset of vasodilatation, which is maintained for longer periods. However, there may be a central nervous or psychological component which permits the subject to ignore the sensations of cold. The author can testify to the efficiency of local acclimatization, and add evidence to show that cooling of the whole body is not necessary to trigger the response. In the southern summer of 1986–87 he carried out experiments in the Antarctic which involved manipulating animals and equipment with his hands several times each day in aquaria held at -1 to $0°C$. The rest of the clothed body was exposed to an equable room temperature. Use of gloves was impossible as fine work was required, and for the first few days it was only possible to tolerate immersion for 1–2 minutes before numbness and pain intervened. Recovery from immersion was even more painful! After a month of such study it was possible to work for much longer periods (< 10–15 min), and recovery was relatively painless. Tactile discrimination was also much improved after this time. Acclimatization was gradually lost within a few weeks of the end of the investigation.

8.4 DAMAGE BY COLD

Hypothermia is the most dangerous manifestation of cold damage in humans. Discomfort starts when the shell temperature falls and the metabolic rate rises with the onset of shivering (Folk, 1969). Hypothermia is said to start when the core body temperature drops to $35°C$ or below (Figure 8.4). However, there are signs of deteriorating condition even before this temperature is reached. Shivering becomes intense, but then begins to die away as body temperature falls to $35°C$ (Edholm, 1978). Some change in behaviour (manifest either as confusion, or unnatural quietness), plus extreme pallor and wheezing (due to pulmonary oedema) are usually noticeable at this stage. As the core temperature falls below $35°C$ muscular weakness, poor co-ordination and impaired breathing rhythm follow. Visual co-ordination fails at a core temperature of $34°C$ (Edholm and Weiner, 1981) and shivering ceases altogether at $33°C$ (Whittingham, 1965), so loss of temperature accelerates. Consciousness is impaired at $32°C$ and lost at $30°C$. Death is almost inevitable as soon as core temperatures fall to 25–$28°C$ (Ward, 1975) because atrial and ventricular fibrillation become likely, causing cardiac function to cease abruptly. Hypothermia of this degree is extremely dangerous, though there have been instances of recovery from core temperatures as low as $20°C$ after immersion in water

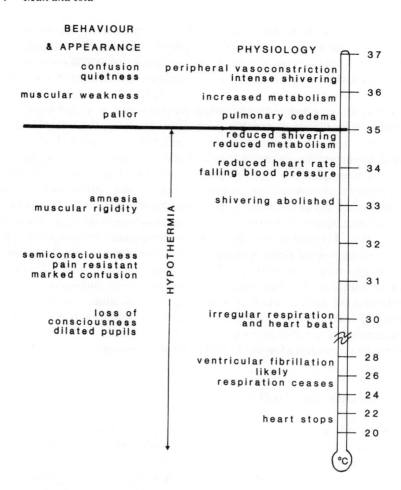

Figure 8.4 Progress of hypothermia in humans.

(Keatinge, 1969). The heart finally stops beating at a core temperature of 18–20°C.

There are many causes of hypothermia. Walking or climbing in cold/wet/windy weather is a common cause, even in the young, fit and well equipped. Edholm (1978) reported a 40–minute progression from walking with support to death in a 15 year-old boy on an autumn hill walk in Scotland. He also commented that a person in rain-soaked clothing exposed to a 16 kph wind might just as well be naked as far as insulation is concerned — this stresses the importance of waterproof and windproof clothing.

Immersion in cold water is particularly potent. Survival times in water of 5°C (common in temperate lakes and seas during winter) may be measured in minutes (panic and shock contribute to this), while victims have survived for several hours in water at 15–20°C. Size, body type and age affect these values. Thin people are far more vulnerable to cold than fat individuals; women are more resistant on average than men. Children can easily become hypothermic in 30 minutes at a water temperature of 20°C, and babies (which have imperfect temperature control anyway) are even more vulnerable.

Hypothermia is also an important cause of mortality in babies and old people living in underheated accommodation — their lack of mobility prevents effective muscular thermogenesis and both groups have poor temperature control — undeveloped in infants, sluggish in the old. Certain medical conditions make the sufferers vulnerable; individuals with Downs' Syndrome are prone to hypothermia when young, as are people with an underactive thyroid gland. Paraplegic patients are also vulnerable to cold as they have substantially impaired ability to produce heat by shivering. Hypothermia often results from excessive drinking of alcohol or drug abuse, either of which may promote heat loss through dilated peripheral blood vessels, and inhibit muscular thermogenesis. It has been said that a high proportion of hypothermic deaths in people working in polar regions have involved drunkenness. High altitude climbers are also at great risk of hypothermia, not simply because they are in a cold environment, but because the hypoxia associated with low levels of atmospheric oxygen prevents effective deployment of normal muscular thermogenesis (Marsigny *et al.*, 1988).

Treatment of mild hypothermia is relatively simple and involves transferring the victim to a warm environment and adding extra clothes and blankets. Sharing a sleeping bag with a warm individual is also recommended in emergency situations. Treatment of severe hypothermia is much more difficult and should ideally be handled by experienced medical personnel equipped with defibrillators and cardiac monitoring gear. Treatment of profound hypothermia has included delivery of warm air to the lungs, peritoneal dialysis with warm fluid (43–44°C) and passage of blood through a warmed external circuit (see Marsigny *et al.*, 1988 for discussion). Paramedic teams dealing with avalanche victims now carry equipment to deliver warmed air to the lungs. First aid manuals stress that external heat must not be applied under any circumstances, nor must the body be rubbed since either of these operations tends to stimulate the peripheral blood circulation, thereby delivering cold blood to an already cooling core. Clearly hypothermia is to be avoided at all costs. Adequate clothing (including waterproof clothing if out of doors) and food is necessary, and during outdoor activties, leaders/organizers must always recognize the dangers and gear their activities to the slowest and weakest participants not to the fastest and strongest. They must always make conservative decisions, cancelling activities if weather forecasts are adverse and responding to deteriorating conditions by immediately seeking shelter or staying put, not by pressing on and hurrying their charges (who may thereby use more energy and cool more quickly as they tire). Dive leaders need to take particular care with cold weather SCUBA operations — a cold, wet diver is at particular risk when transferred to land by small boat. Unless protected from wind and the speed of the vessel he or she may deteriorate rapidly. Recovery from hypothermia is usually uneventful, but complications have involved lung infections stemming from pulmonary oedema (i.e. swelling or waterlogging of lung tissue), plus kidney damage produced by low temperatures and hyperviscous blood.

Other damaging effects of cold are more peripheral and less immediately life-threatening. They range from minor skin damage (chapping and chilblains) to trenchfoot and frostbite. Trenchfoot has largely been identified in war time, particularly during the two world wars. During the First World War it was seen in troops who had their feet immersed in cold water in the trenches for long periods (hence the name). It appeared again in the Second World War, but mainly amongst survivors of torpedoed ships whose feet had dangled in the cold sea for many hours or days. In both cases the peripheral nerves and blood vessels were damaged, often permanently, by prolonged local hypothermia. Probably some of these effects derived from increased skin permeability to ions rather than temperature *per se*, but the final result was often gangrene and death.

Frostbite is an ever-present risk at temperatures substantially below 0°C. Essentially it is the freezing of peripheral tissues, usually of the extremities. Amongst native peoples (e.g. Inuit, Saami) and experienced polar workers frostbite is extremely rare because they dress adequately and seek shelter in poor weather. They also recognise the symptoms of frostbite (white areas on extremities and face) in their colleagues and treat such mild cases before significant damage is caused. Frostbite tends to afflict drunks and victims of accidents or violence. Polar explorers in the days of Amundsen and Scott were vulnerable because of their extreme situation and generally poor nutrition (scurvy was common) and equipment. Modern Himalayan mountaineers are still at risk because of the low temperatures, relatively poor mobility of limbs (because of the difficulty of the terrain and low oxygen tensions) and inadequate shelter. Frostbite damages tissues by osmotic dehydration (as the osmolarity of unfrozen extracellular fluids rises) rather than by intracellular freezing (Figure 8.5). In severe cases blood vessels are damaged or blocked by necrotic tissue and gangrene is a risk. Treatment involves rapid warming if the frostbite has happened quickly (as may happen within 30 seconds in central Antarctica if flesh is exposed!). However, if the frostbite has been maintained for long periods, slow warming is recommended. Administration of drugs to prevent blood clotting (e.g. heparin, warfarin) and dilate blood vessels are also employed by medical personnel.

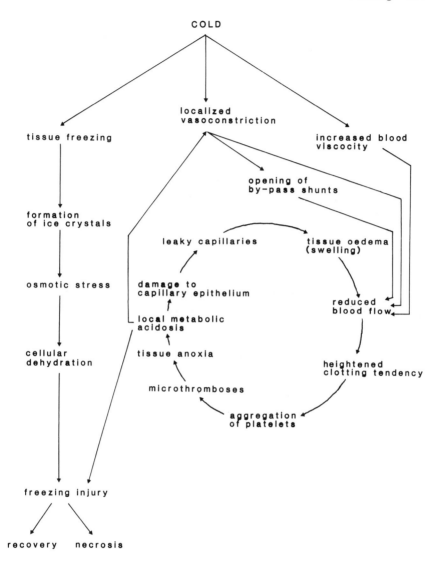

Figure 8.5 Progress of freezing injury ('frostbite') in humans. Redrawn and translated from Marsigny *et al.* (1988).

8.5 CLOTHING

Man in the tropics can survive naked, even without fire or shelter. Progressively thicker and more extensive insulative clothing is needed for survival the further he moves away from equatorial regions (especially in winter). It is clear that man (including *Homo erectus* as well as the later *Homo sapiens*) has employed clothing derived from animal skins for hundreds of thousands of years, and clothing derived from plant fibres for at least 7000 years (and probably much longer). The insulative value of clothing depends upon the amount of air trapped within the fibres/hairs of the clothing, and between different layers of clothing. Preindustrial man adopted a variety of approaches depending on the resources available to him. Inuit, who spent much of their life out on the ice, relied virtually completely on animal skins, sewn together with bone needles and leather thongs or ligamentous/tendinous material taken from the marine mammals and caribou they relied upon. Samic people also used animal skins, especially reindeer skins, but supplemented insulation (particularly of their fur boots) with hay. Hay is a highly effective insulator, provided that it is kept warm and dry. Both groups were aware of of heat loss from the head — Inuit incorporated fur lined hoods into their parkas, while Saami used skin or cloth caps, usually stuffed with straw or hay.

Iceland has been occupied (initially by Irish monks, then by Scandinavian immigrants) for over a thousand years. One of the island's less appreciated natural assets is the large population of Eider ducks (*Somateria mollisima*). For centuries the eiders have been ranched, protected against predators by eider 'shepherds', not for their meat, but for the down feathers which they use to line their nests. Eiderdown provides remarkable insulation, whether used in bedding or in clothing, and took its place amongst animals skins in the apparel of Icelanders. Down is particularly valuable because it is elastic, so recovers well from compression (provided it is kept dry).

When European and American explorers started to visit the polar regions with increasing frequency, they initially simply used several layers of conventional cold climate clothing, mainly made from wool and cotton. Some (particularly the Norwegians) adopted and adapted skin clothing of the Inuit/ Greenlander pattern, and a mixture of these approaches prevailed until the Second World War. During the past 50 years, the approach to clothing has become more scientific, and has also involved increasing use of man-made materials (e.g. nylon and polypropylene fibres, plus perspex for goggles and visors, plastics for boots). Particular attention has been given to the problem of allowing hard exercise in polar regions. Any sort of exercise tends to lead to peripheral vasodilation and sweating. Loss of sweat makes absorbent materials damp, thereby degrading the insulative properties of clothing. When exercise ceases, the wearer loses heat rapidly. This difficulty has been dealt with in two ways. First, the British Antarctic Survey subscribes strongly to the 'layer principle', requiring its staff to wear several layers of thin clothing over all of the body (including head and extremities). During exercise, layers are removed to maintain comfort and minimize sweating; they are replaced when exercise ceases. The Survey believes this approach to be superior to reliance on smaller numbers of thicker garments, even when these contain down.

Second, particular difficulty is involved when people are exposed to snow or rain as well as low temperature. Good waterproof materials have been available for many years (oilskins and waxed cloth were early examples), but their impermeability creates a problem since they prevent the escape of sweat, so that exercising in such garments is most inadvisable in cold climates. During the last few years even this problem has been alleviated to a considerable extent by the development of items of clothing which have a porous plastic membrane beneath the outer cloth. The membrane has pores which permit the passage of water vapour (so allowing the water of perspiration to escape), but the pores are too small to admit the water droplets of rain or melted snow which are on the outer surface of the garment.

The insulative value of clothing together with its trapped air can be described in terms of 'clo' units, the resistance to heat flow (more properly measured in joules m^{-2} deg. C^{-1}) along a temperature gradient between the body surface and the environment. One clo is equivalent to the clothing needed to assure comfort at a room temperature of 20°C (roughly speaking a business suit with the usual underpinnings!). Insulation provided by human peripheral body tissues (assuming vasoconstriction) is about 0.8 clo (Burton and Edholm, 1955) (this compares with clo values of about 5 for the peripheral tissues and fur of polar bears), whereas a standard polar outfit consisting of (1) woollen underwear and socks; (2) woollen long sleeved shirt, moleskin trousers and shoes; (3) coveralls, woollen gloves and cap; (4) parka with hood, mittens and fur-lined boots has an insulation value of 4 clo (i.e. 5 times natural insulation) (ASHRAE, 1981). A wind of $15+$ m s^{-1} will reduce insulation of 'windproof' clothing by about 30%; dampening of clothes by sweat can remove 25%. It is important to know such information, because it allows calculation of sensible working periods in polar climates. De Freitas and Symon (1967) have used data about clothing characteristics, metabolic rates under different work regimes, environmental temperatures and wind speeds to construct a bioclimatic index of survival times in the Antarctic. Indefinite survival is taken to be possible under those conditions which allow an individual to maintain a constant core body temperature of 37°C indefinitely (if in some discomfort!). Large areas of the central Antarctic are lethal throughout the year. This means that core temperatures will inevitably fall if the individual remains out of doors. The most severe weather conditions are life threatening within 20 minutes and lethal within half an hour. Even at the tip of the Antarctic Peninsula in summer, indefinite survival requires full polar clothing as described above; it is salutary to remember that much fiercer conditions occur in upland areas of the UK in winter!

The head and extremities are difficult to insulate successfully with clothing. Insulated helmets with visors have proved effective for skidoo driving at high latitude (an activity which obviously guarantees windchill!), but insulation of the hands is a particular problem. Thick gloves prevent fine work, and in any case often reduce insulation rather than improving it! The reason for this apparent anomaly is that cylindrical insulation around cylindrical objects rapidly increases the surface area for heat exchange. Obviously there is a trade-off with insulation thickness, but a more effective approach is to combine thin gloves with substantial overmittens (with separate thumb) which can be removed for short periods to permit fine work.

In the past 40 years an increasing number of people have become involved in an activity which routinely exposes them to cold — diving using self contained underwater breathing apparatus (SCUBA). Except in warm tropical waters above 27°C this recreational activity involves the potential for rapid heat loss, since water has such a great heat capacity (Chapter 1). Over the years, much progress has been made in devising insulative clothing. Most SCUBA divers in the 1960s and 1970s used wet suits made of thin neoprene rubber. This trapped water warmed by body heat against the skin (which reduced the discomfort of cold) and, since the neoprene is filled with air bubbles, also provided insulation, though it has to be said that wet suits delay cooling rather than preventing it. Thicker suits were used in colder waters, but the practical maximum thickness was found to be around 8 mm — thicker suits reduced joint mobility and hence swimming power. Such suits permit dives of 30–60 min in water of 2°C. In recent years semi dry suits have become popular (these avoid the flooding with cold water that afflicts the wearers of wet suits when they change swimming attitude), but for regular cold diving, most exponents of the sport choose dry suits. Dry suits are made of tough sheet synthetic rubber, are sealed at neck, ankles and wrists, and allow the diver to wear ordinary clothing underneath (though most divers now wear purpose-designed fleece one-piece 'woolly bears'. Insulation of the head has been difficult, particularly the forehead which has usually been exposed to ambient temperatures. Commercial professional divers now wear enclosed helmets, but divers of the British Antarctic Survey were still using wet suits and diving in water of −1.9°C with exposed faces in the late 1980s!

If work is to be done at very low temperatures for long periods (e.g. by pilots flying at extreme altitudes, astronauts moving outside their space vehicles, deep-sea divers working in the arctic at depths requiring decompression), then clothing must involve built in heating, either electrically or, in the case of divers, by the pumping of warm water through their suiting.

8.6 SHELTER

Shelter is a vague term, but encompasses everything from a windbreak or snow hole, to a mansion! We have already seen (Chapter 3) that nest or shelter building is widespread in the animal kingdom. This is true of primates too, and all of the great apes build nests. These have been particularly studied by

Goodall (1971) in chimpanzees (*Pan troglydytes*), which are believed from modern genetic evidence to be the closest living relatives of *Homo sapiens*. Chimpanzees often build complicated nests with much interweaving of branches, twigs and leaves, though they may be satisfied with a few twigs on other occasions. Shelter of this sort may provide privacy, camouflage or have social functions, but the building of simple vegetation shelters would presumably have been a simple matter for early men as they moved into subtropical and temperate areas (initially much more densely wooded than today), encountering progressively cooler nights. Tool usage would have helped in the construction of huts, and with such shelters comes protection from windchill and the conservation of heat which rises from the body.

Huts and gregariousness must have been of great importance to the first men who moved out of their thermoneutral zone. Much has been made of cave dwelling habits in men living perhaps 25 000–100 000 years ago in Europe, but caves are relatively rare and reliance on them throughout the year would reduce flexibility of movement. It seems likely that most nomadic hunter-gatherers lived in temporary huts of wood and leaves or reeds. Even the caves may well have had light structures at the entrances to protect against windchill. One of the most striking pieces of evidence of the antiquity of shelter building was provided by the discovery of an archaeological/palaeontological site at Nice, on the shore of the Mediterranean (de Lumley, reported in Leakey, 1981). The site was repeatedly occupied by people (*Homo erectus*) about 400 000 years ago. At this time the northern coast of the Mediterranean was cool temperate in nature, with heather, pine trees and oaks covering the slopes around the site. It appears that one or more groups of hunter-gatherers regularly visited a sheltered pebbly beach each spring for a number of years. They build simple huts by erecting a row of stout posts, then leaning branches against the row. The posts and branches were stablized against wind by big stones, and the shelters were quite large (12×6 metres) so would have accommodated several people. Whilst there they ate red deer, elephant, rhinoceros, wild boar and goat from the surrounding woods; they consumed mussels and limpets from the shore (the remains of all of these food items being fossilized). They also used animal skins to sit or lie upon and had simple flint tools.

Penetration of higher latitudes involved more substantial housing. The remarkable insulation and careful architecture of Inuit igloos has enthralled many commentators, but the turf houses of the Samic peoples (beautifully displayed in the Arctic Museum of Tromsø, northern Norway) are equally impressive and became elaborate in the 19th century. Perhaps the most bizarre of arctic habitations were the huts made of leather stretched across arched mammoth bones — building materials employed in the Ukraine and Poland 21 000–23 000 years ago and which involved the slaughter of dozens of mammoths (though presumably primarily for food rather than architecture!).

Caves, igloos, turf houses and mammoth-frame huts were probably the winter bases of nomadic communities who continued to use more flimsy shelters in spring, summer and autumn as they followed the herds of their prey (or semi–domesticated reindeer herds in the case of the Saami). Until relatively recently Inuit and Samic peoples still used wigwams or teepees made from a few poles and animal skins when moving around in the summer, shelters little different from those employed throughout the year by groups living in more moderate climates (e.g. Amerindians).

8.7 FIRE

The discovery of methods of making fire for warmth, scaring animals and for cooking occurred several hundred thousands of years ago. *Homo erectus*, the dominant hominid before the appearance of *Homo sapiens*, undoubtedly used fire in northern China 500 000 years ago, since layers of ash have been found in the caves of Choukoutien where 'Peking Man' was first discovered in 1927 by Davidson Black (see Leakey, 1981 for review). How this came about is un- clear. Natural sources of fire include lightning and volcanic activity, the former widespread but ephemeral, the latter localized but often reliable over long periods. Initiation of fire (usually by friction or flint-striking before the advent of matches) and the maintenance of fuel supply are the main problems involved in using fire. Many methods of transporting fire appear to have been developed by preindustrial man.

Cooking is widespread in the tropics, and it is probable that fire was em- ployed long before man moved into cooler latitudes. Perhaps fire was first used to keep great carnivores at bay, maybe it was used to soften meat — we shall never know. Fire and its possession have had great cultural significance, par- ticularly for men living in cold climates where it provides warmth and comfort, permits utilization of a greater variety of foods than could be consumed raw, and also contributes to simple high latitude technology — in particular it allows the extraction of oil from blubber, thereby yielding light, and the waterproofing of leather. Hot cooked food also provides rapid heat input to the body's core — as important to the elderly living in underheated accommoda- tion today, as it was to the Canadian trapper of the last century.

8.8 COLD AND HUMAN DIET

Man's closest living relatives are largely vegetarian, though chimpanzees do take some meat when they can get it — a variety of monkeys are caught by organized hunting, while individual chimpanzees use simple tools (sticks) to collect insects. !Kung tribesmen and women are amongst the few remaining tropical hunters and gatherers. The women provide some 75% of the diet by gathering roots, vegetables and berries; the men contribute 25% of the diet in the form of meat. Interestingly, both sexes need to work about 40 hours per week to achieve this distribution. As man moved into colder climates, it appears that the proportion of meat/fish in his diet rose, probably for two reasons. First, edible roots, vegetables and berries become increasingly seasonal in their availability with increasing latitude. Even as little as half a century ago people in the UK would expect to rely on stored potatoes, carrots, turnips and apples for their winter plant food — and such storage did not happen until man abandoned a nomadic lifestyle. Early man living in cool temperate zones would subsist mainly on meat and fish in the winter, probably following the herds of caribou and mammoths, but would push up his plant intake in the spring and summer. A 50% plant : 50% meat diet was probably representative of this sort of life (averaged annually).

Second, the increased metabolic demands of life in cold climates necessitated increased food intake. Meat and fish, particularly the fat component, provide more energy than plant material. There is evidence that increased meat eating (helped by improved technology of prey capture) may have contributed to increased leisure time, and hence to more complicated rituals and culture.

For Inuit and Saami these two factors interacted to ensure that the proportion of meat and fish in the diet reached very high levels indeed before these peoples were influenced by southerners. Inuit in particular probably relied on fish, seal, walrus, polar bear and caribou for well over 90% of their yearly diet, with these items (which could be naturally stored in the deep-freeze of their winter environment) being their sole energy source for most of the year. A particularly noteworthy feature of the Inuit diet was the extremely high lipid component — important in terms of energy supply, but distinctly unpalatable to the Europeans who first encountered the Inuit! At one time it was not understood how the Inuit could eat a high lipid diet, yet not have a high incidence of cardiovascular disease; it is now appreciated that the lipid intake is predominantly of unsaturated fatty acids, and includes arachodonic acid, now thought to provide positive protection against heart disease.

Recent archaeological/palaeontological research suggests that interaction of cold, shelter and diet may have had much to do with the development of fixed communal living (in contrast to the roving bands of hunter-gathers, each probably made up of no more than 30 people). Traditionally this has been linked with the development of pastoralism and agriculture in North Africa, the Middle East and Central America some 5000–10 000 years ago, but it is becoming clearer that fishing in particular becomes more efficient when operated by quite large groups operating from fixed bases, and that fishing 'villages' may have existed in northern Europe for tens of thousands of years, perhaps with associated management of plant resources. There is also evidence that groups of early *Homo sapiens* may have coralled horses near permanently occupied caves and seem to have developed the use of reindeer and horses as beasts of burden (and for milk?) much earlier than once recognized.

8.9 MEDICAL HYPOTHERMIA

Until the 1950s two organs were largely inaccessible to surgery; the heart and the brain. Both organs require good supplies of oxygen to function at normal body temperature. Deliberate cooling of patients to core body temperatures of 26–28°C permitted increasingly complex cardiac surgery, and some brain surgery has involved cooling to 10°C; these temperatures greatly reduce oxygen consumption and prolong the period during which blood supply to the organs can be reduced or inhibited. During medical hypothermia the natural behavioural reactions to counter low core temperature and involuntary muscular thermogenesis have to be overcome. This is achieved by a combination of general anaesthesia and use of neuromuscular blocking agents (such as curare). Recovery from medical hypothermia is achieved by delivery of warm gases (air and anaesthetic) to the lungs, warming the blood via external circuits and defibrillation of the heart by application of strong electric impulses.

Maintenance of tissues/organs at low temperatures is also a necessary adjunct of transplant surgery. Kidneys in particular often have to be transported over long distances between donor and recipient; maintenance of low temperature (but above freezing) sustains good tissue condition and makes successful transplantation more likely.

A specialized low temperature technology in medicine has involved the use of extremely low temperatures to destroy tissues with minimal release of cells into the circulation ('cryosurgery'). Cryosurgery has been used particularly upon malignant growths in inaccessible organs (e.g. brain tumours) and has involved implantation of insulated metal electrodes, which are then cooled quickly with liquid nitrogen. At a more superficial level, warts and skin tumours have been destroyed with the aid of freezing sprays and application of liquid nitrogen.

8.10 COLD TECHNOLOGY

The efficiency of low temperature as an agent of preservation has been appreciated for centuries, but has come to be applied for domestic, industrial, medical and scientific purposes only in the past 200 years. In the late 18th and early 19th century many large European houses had 'icehouses', well insulated buildings in which ice was stored from the winter until the summer. An example may be seen in Penrhyn Castle (near Bangor in North Wales). In this instance nearby fields were deliberately flooded in cold winter weather to allow the formation of large quantities of ice, which were collected and stored in a large, windowless, cylindrical room with thick stone walls which was built on the shaded side of the mansion. Blocks of ice would then permit safe storage of raw meat for periods of some days, even in summer weather.

In the later part of the nineteenth century, the discoveries of methods of extracting heat by evaporative cooling, and the liquifaction of gases by the application of great pressures progressively led to the refrigeration technologies that we now tend to take for granted, but which are crucial to much of the modern production, transport and marketing of food, as well as to the storage of medical products.

The technology of preservation of live biological material at very low temperatures is often referred to as cryopreservation. Its basis is the virtual abolition of metabolism of cells and tissues held at extremely low temperatures (usually $-196°C$, the boiling point of liquid nitrogen). However, protocols of freezing demand that cells are exposed to minimal likelihood of osmotic dehydration and intracellular freezing. Careful control of cooling rate is required, plus the presence of organic cryoprotective agents. Preservation of red blood cells (erythrocytes) with glycerol was pioneered in the 1940s, this technique opening up the possibility of preserving large quantities of blood cells for transfusion (with blood plasma) during war time. High rates of cooling were required $(1000–5000 \text{ deg. C. min}^{-1})$. Preservation of bovine spermatozoa followed shortly afterwards, allowing the rapid introduction of artificial insemination programmes. Further progress was made with the use of dimethyl sulphoxide (Me_2SO) as a cryoprotectant which is non-toxic, yet penetrates plasma membranes. By and large freezing and thawing protocols have been devised by trial and error — a 'recipe' approach (Franks, 1985). A variety of sugars, polyols and man-made polymers have been used to minimize damage during thawing, but this has been done on a pragmatic basis with little underlying theory. Lehn-Jensen (1986) gives a good account of the research necessary to devise protocols for the successful cryopreservation of bovine embryos, and the genetic benefits to be gained by the transfer of embryos derived from 'high quality' parents, to the uteri of 'low quality' cows. Cryoprotective concentrations are important, but so are cooling rates; too slow a cooling rate allows harmful osmotic gradients to be built up, too fast a cooling rate causes lethal intracellular ice formation.

Although notable successes have included the long term storage of sperm, cattle zygotes and even human embryos (with $< 80\%$ survival in the last case), it must be appreciated that all of these have involved 'freezing' of single cells or small cell clusters (e.g. bovine and human embryos). Cryopreservation of organs or whole multicellular animals has so far proved impossible, because of the formation of damaging thermal, osmotic and ionic gradients within such larger structures.

Although cryopreservation is increasingly important in agribusiness and medicine, we are still far away from that cryopreservation of whole humans, beloved by science fiction writers as a tool for the conquest of space and time!

PART FIVE

Cold and Evolution

9 Evolution and low temperature

9.1 GENERAL CONSIDERATIONS

For evolution to take place in a population of animals there must be genetic variability within that population and selection pressures (biotic or environmental) to act upon that variability. The variety of genetic material must be expressed in some preadaptive feature(s) of form, function or behaviour. There is little difficulty in the *a posteriori* acceptance of the importance of low environmental temperature as a selective agent which drives evolution, indeed this book is full of examples of adaptation to cold conditions in which such selection is implicit. However, selective pressure is only one part of the equation, genetic variability being the other. If low temperature acted directly to enhance variability, evolutionary processes could be stimulated.

Genetic variability stems from two sources; mutation and recombination. During reproduction, genetic information is transmitted from one generation of animals to the next in the form of DNA macromolecules. A mutation occurs when the sequence or number of nucleotides in DNA is altered in the parent and the novel arrangement is passed onto the offspring. Mutations may affect a small part of a single molecule of DNA ('point mutations'). Alternatively, they may change large parts of a chromosome (each chromosome is believed to contain a single DNA molecule), or even the chromosome number ('chromosomal mutations'). The term 'mutation' therefore covers a wide range of alterations in genetic material (see Goodenough, 1978 for a good introduction). However, there are common features to mutations. All may occur spontaneously, yet can be induced by physical factors or chemical agents. Mutations involve drastic changes to genetic material, which are therefore likely to have profound effects on a species' biochemistry. It is therefore unsurprising that most mutations are lethal or deleterious. However, the tiny fraction of favourable mutations provide a basis for evolutionary change (the only one in asexual species).

Recombination is a much gentler and more common agent for the production of novel genetic combinations. During the sexual process (sexual in the widest sense of the word), homologous chromosomes line up with one another, gene for gene. The chromosomes then exchange pieces of DNA so that neither shows a net loss or gain of nucleic acid. The possible combinations of nucleotide sequences which result are enormous, particularly in species with numerous chromosomes, so a variety of genotypes are produced on which selection may act. The frequency of exchange of material between chromosomes is known as the recombination rate; the higher the recombination rate, the greater the variety of genotypes produced.

It is evident that any environmental factor which might alter mutation or recombination rates will influence variability directly (and, hence, selection indirectly). Is there evidence that low temperatures have such effects? In answering this question we are handicapped by the narrow range of organisms studied by geneticists; fruit flies and mice still dominate! As a generalization, rates of biological development increase with temperature, so *a priori* one might expect mutation and recombination rates to fall at low temperatures (at least for ectotherms). However, Lindgren (1972) demonstrated greatly enhanced mutation rates in *Drosophila melanogaster* exposed to extreme temperatures (whether high or low) or to heat or cold shocks. Much earlier studies (Plough, 1917; Stern, 1926; Mather, 1939) demonstrated that recombination in fruitflies was stimulated by low or high temperatures, while Zhuchenko *et al.* (1985) have recently shown greatly increased recombination in *Drosophila* exposed to extreme temperature fluctuations (from 15–29°C). Clearly, low temperature can stimulate mutation and recombination rates, but how widespread this phenomenon is in the animal kingdom is unknown. The basis of enhanced mutation and recombination at low temperature has been attributed to disturbance of intracellular homeostasis (Kergis, 1975; Belyaev and Borodin, 1982; Parsons, 1987), though it may also be suggested, at least in the case of mutations, that genetic repair mechanisms are perhaps less effective at low temperature. If mutation and recombination are stimulated by divergence from a physiological optimum, then it can be argued that low temperatures will be particularly effective in enhancing variability in endotherms, whose physiological optima occur at fairly high temperatures.

A particularly sudden and extreme form of genetic change is that of polyploidy (multiplication of chromosome number). Common in plants, it has increasingly been recognized as an important feature of evolution in animals too. Because birds and mammals have more DNA per cell than do fish and other chordates, and because they exhibit many duplicated gene loci, it is believed that genome doubling (tetraploidy) has taken place one or more times in vertebrate evolution (Ohno, 1967). Polyploid amphibians and reptiles are quite common (Bogart, 1980), while there are a number of naturally occurring polyploid fish, some apparently tetraploid, others triploid (see Allendorf and Thorgaard (1984) for review). Triploidy stems from failure of reduction division in the production of sperm or eggs. Instead of two haploid gametes fusing to yield a diploid zygote, a haploid and a diploid gamete may fuse to yield a triploid animal. Triploids, even if they survive, are usually sterile, but if they can reproduce they reproduce parthenogenetically, the eggs being triploid. Triploidy is often associated with hybridization between closely related species.

Tetraploids could obviously be produced by the fusion of two diploid gametes, but persistence of a bisexual tetraploid strain would require the independent production of a tetraploid male and a tetraploid female which subsequently mated. An alternative to this low probability scenario has been proposed by Schultz (1969) and Astaurov (1969). They suggest that a tetraploid strain could be produced in two steps. First a unisexual triploid strain could arise and reproduce parthenogenetically in the manner described above. Next, members of the triploid strain could occasionally be fertilized by normal diploids, yielding fertile tetraploid offspring (in each case by fusion of one triploid egg with a haploid sperm). Initially tetraploids have four replicas of each chromosome, but over time there is a tendency towards restoration of disomic inheritance ('diploidization').

Higher levels of ploidy than tetraploids appear not to be known amongst vertebrates (except in the strict sense where tetraploidy may have happened more than once in a lineage). However, in some invertebrates remarkably high levels of polyploidy have been recorded. For example, the small intertidal clonal bivalve mollusc, *Lasaea rubra* has been shown to occur with several chromosome numbers corresponding to triploidy, pentaploidy and hexaploidy (Thiriot-Quiévreux *et al.*, 1988, 1989; Tyler-Walters, in press).

Polyploidy has a number of consequences. First, to accommodate the extra DNA, there has to be an increase in cell size, and it is believed that this prevents viable polyploidy in animals with a high degree of tissue complexity (White, 1973; Niebuhr, 1974). In consequence, polyploidy in mammals (including Man) and birds seems to be invariably lethal. Second, the increase in amount of genetic material creates the potential for greater allelic variation, at least in bisexual tetraploid species (for a more detailed discussion, consult Allendorf and Thorgaard, 1984). Third, the presence of extra gene copies slows responses to selection for recessive alleles (but makes little difference to selection for dominant alleles). In summary, polyploidy can produce new species suddenly, which thereafter tend to change relatively slowly. Increased cell size is associated with a reduction in metabolic rate and a slowing of development, so polyploids usually move towards k-selected life history strategies. In salmonid fish, which are now believed to be polyploid, there is a tendency towards the production of few, very large eggs which develop slowly.

The bivalve genus *Lasaea* provides an even better example; the diploid *L. australis* releases large numbers of planktonic larvae, while the self-fertilizing hermaphrodite polyploid *L. rubra* broods a small number of large embryos which are released as miniature crawl-away adults (Tyler-Walters and Crisp, 1989).

Of course we cannot tell what environmental factor triggered polyploidy in a particular species some thousands or millions of years ago. However, because polyploid animals, particularly fish and shellfish, have desirable characteristics from an aquaculture point of view (they usually show poor gonadal development, so more energy is directed towards putting on flesh), there has been much interest in the artificial induction of polyploidy. In many cases chemical induction has been applied, but thermal shock, especially cold shock, has often been found effective (e.g. Lincoln, 1981; Reftsie *et al.*, 1982). It therefore seems likely that a climatic shift towards cold and changeable conditions will increase the frequency of induction of natural polyploidy.

Evolutionary change is known to take place more rapidly in isolated populations of animals. Indeed, at least partial isolation is crucial to allopatric (or geographical) speciation which underpins the concept of 'punctuated equilibria' introduced by Eldredge and Gould (1972) to explain the discontinuous nature of the fossil record and the rarity of 'missing link' species. Isolation prevents gene flow from the bulk of the ancestral population, and this is probably a major factor in promoting allopatric speciation. However, competition in isolated ecological communities is likely to be fiercer, favouring niche specialization and speciation. Much evidence has been derived from island and lake material (e.g. the finches of the Galapagos Islands, dwarf fossil elephants and hippopotamuses on Mediterranean islands, cichlid fish of Lake Victoria), so it may be argued that warm conditions, which raise sea levels and isolate bodies of land to a greater or lesser extent, tend to cause isolation of populations of terrestrial and freshwater animals, hence providing the right circumstances for the rapid selection of favourable characteristics. However, low temperatures may also isolate groups of animals. During periods of glaciation, bodies of shallow sea may be separated from one another as sea levels fall, cutting off genetic communication between populations of fish and invertebrates. Hills and mountains, formerly wooded and hospitable to animal life, become frozen, barren barriers. Glaciers descending from icecaps to the sea can create refugia between them (Lindroth, 1969; 1972) where plants and animals can survive (and become modified in their low temperature isolation). Colonization of deglaciated areas by a diverse beetle fauna derived from such refugia is believed to have taken place in western Europe after the last Ice Age (Coope, 1969).

Finally, changes of world climate have profound general effects on environments, with the effects being greater at higher latitudes. A shift to low and changeable temperatures tends to result in a reduced diversity of vegetation and a loss of ecological niches. This niche loss is likely to intensify inter- and intraspecific competition, while the changes in temperature will impose novel section pressures on animal populations. For geographical reasons there will again be a tendency for populations at the high latitude edges of species' distributions to be affected most (favouring allopatric speciation).

In summary, therefore, low temperature not only acts as a selective agent, but also has direct effects on genetic material and the environment which are themselves likely to increase genetic diversity and to speed up the action of selection.

9.2 COLD AND THE EVOLUTION OF ENDOTHERMY

Broadly speaking, endothermic animals have two advantages over ectotherms. First, they can maintain relatively constant body temperatures in both cold and hot environments, thus permitting animals to remain active through most of the day and night without wasting time in basking and shade seeking. Second, the high metabolic rates required to sustain endothermy permit endotherms to display high levels of activity over long periods. It seems inconceivable that high metabolic rate preceded physiological control of temperature, since such a rate would be enormously wasteful if the animal concerned did not have effective insulation and control over peripheral blood flow. Many small living reptiles (snakes and lizards) have physiological mechanisms to control heat uptake and loss (Heath, 1964; Bartholomew and Lasiewski, 1965; Morgareidge and White, 1969) despite their low metabolic rates. Although these animals do not possess insulation, they are capable of controlling the distribution of blood flow. This allows them to divert blood to the periphery during bouts of basking in the sun (thus facilitating rapid warming of the whole body), yet they shunt blood centrally and constrict superficial vessels when exposed to cold conditions, so that loss of heat is slowed. It is easy to see how a reptile with control over its blood distribution could progressively become endothermic, if it first evolved insulation (low thermal conductance) and then achieved a heightened metabolic rate (by increased aerobic muscle bulk and perhaps brown fat). Cold conditions encountered primarily at night, but perhaps also on a seasonal basis, presumably provided the selective pressure which led to the evolution of the earliest endotherms.

We have already seen that small endothermic birds and mammals are rather specialized and require relatively huge amounts of food, so it is probable that endothermy first evolved in fairly large animals. McNab (1978) suggested that a body weight of 30–100 kg was appropriate; in this he was influenced by studies on the Komodo dragon, *Varanus komodoensis*, the giant monitor lizard of Indonesia which possesses a degree of 'inertial homoiothermy' by virtue of its body size (McNab and Auffenberg, 1976). Inertial homiothermy describes a state in which an animal maintains a fairly constant body temperature simply because it is large, and not because of a heightened metabolic rate. A large reptile (like a Komodo dragon or adult alligator) takes a long time to warm up or cool down because it has a low surface area to volume ratio. Given a subtropical climate with hot days and cool/warm nights, such animals can maintain remarkably steady temperatures. McNab (1978) suggested that large reptiles evolved fur (which improved their inertial homoiothermic capabilities), and that these furry precursors of mammals led to smaller species with higher weight specific metabolic rates — true endotherms. McNab provides no basis for the reduction in size, or reasons for the elevated metabolic rates, but was undoubtedly trying to explain why the earliest known mammals were small (and probably nocturnal).

This leads us to one of the most fascinating paleontological debates of recent years, concerned with when endothermy arose, and in what group(s) of animals. Half a century ago there was little room for doubt in this matter. Only two groups of animals were recognized as endothermic (birds and mammals); the 'lower' mammals (monotremes and marsupials) had 'poor thermoregulation', while reptiles were cold blooded ectotherms. It was clear that endothermy had evolved twice and could only have existed in the immediate precursors of birds and mammals. This was the era when consciousness of a hierarchy of abilities within the vertebrates was very strong; amphibia were 'better' than fish, reptiles more 'advanced' than amphibia and so on. This approach is unfortunately still pervasive (probably because of the long publishing runs of some textbooks!), and undergraduates still sometimes need convincing that all species were and are fit for their environment and time.

Progressively it has been realized that endothermy is more widespread than previously thought (leatherback turtles, some female pythons and many flying insects are endothermic during part of their life), that monotremes and marsupials have good thermoregulation, and that many mammals and birds (e.g. camels, bats, hibernating rodents, penguins and humming birds) may be partially ectothermic under certain conditions, for good adaptive reasons!

In 1971 Robert Bakker started a controversy which has been raging ever since (see Bakker, 1986 for a recent and entertaining summary). Put simply, he believes that the bulk of the dinosaurs were high metabolic rate endotherms, that birds are effectively feathered dinosaurs, and that mammals (which existed as small nocturnal animals throughout most of the period of dinosaur domination) were specialized furry derivatives of larger efficient reptilian endotherms. Bakker's evidence has been drawn from comparative anatomy (including some telling histological evidence provided by de Ricqles (1974) from observations on dinosaur bone structure), from deductions about the life styles of the various herbivorous and carnivorous dinosaurs, from assessment of prey-predator ratios in living and fossil animal assemblages, and from calculations of speeds of dinosaur locomotion (the last being derived from

fossil dinosaur tracks). Bakker also suggested that the pterosaurs were endotherms, relying here on the observation of Sharov (1971) that at least some
pterosaurs possessed fur on their bodies. In the author's view, Bakker's case is
very strong, especially as it provides good reasons for the suppression of
mammalian adaptive radiation during the millions of years of dinosaur domination. Since endothermy allows much higher levels of sustained activity than
does ectothermy, it seems inconceivable that mammals could not have evolved
larger size and outcompeted dinosaurs — unless the latter were themselves
effective high metabolic rate endotherms.

The concept of the endothermic dinosaur has not been universally accepted.
Part of the problem lies in the common assumption that dinosaurs inhabited a
universally warm world which only became cool at night. While it is true that
tropical conditions probably extended to latitudes as high as 45° during most
of the Cretaceous, dinosaur fossils have been found at sites (e.g. in South
Australia) which were at 70°S at the time of the animals' death (remember that
Continental Drift has moved continents considerably!), and the associated
vegetation and geochemistry indicate that the climate was temperate, with
winters cold enough to freeze lakes. Opponents of Bakker's hypotheses have
concentrated on the idea of inertial homoiothermy, demonstrating that large
reptiles could maintain constant body temperatures despite diurnal temperature changes, whilst still having an ectothermic physiology. Spotila *et al.*
(1973) produced a mathematical model (still much quoted) that demonstrated
that a large reptile (with the general physiological characteristics of an alligator)
could maintain a near constant body temperature in modern Florida during
the summer throughout the diurnal cycles of temperatures, especially if the
animal had an insulating layer of fat. However, the model did show that
such an animal would cool down within a matter of days during prolonged cool
weather. It is unfortunate that modellers who have considered dinosaurs have
relied so much on crocodilians for their metabolic rates. Alligators and crocodiles are specialized tropical and subtropical *aquatic* sit-and-wait predators
which catch most of their prey at the air-water interface throughout their life
(Davenport and Sayer, 1989; Davenport *et al.*, 1990a). They spend most of
their time motionless and have extremely low average levels of metabolism as
adults — Cott (1961) calculated that an adult Nile crocodile (*Crocodylus
niloticus*) ate about the same quantity of food in a year as a pelican! Their
stomachs are often empty and they assimilate energy and protein with great
efficiency from infrequent meals. They are hardly comparable with large terrestrial herbivores and carnivores which could not wait for food to come to
them. Spotila *et al* and other workers have also concentrated upon the problems of the very large sauropods (e.g. *Diplodocus, Brontosaurus (Apatosaurus)*).
Some sauropods were as large as 20–30 tons, perhaps 5–8 times as heavy as a
modern African elephant (*Loxodonta africana*). It is well known that an African
elephant has something of an overheating problem when exposed to the force

of the African sun, only soluble by ear flapping, shade seeking and evaporation of water sprayed over the body by the trunk. Modellers have suggested that sauropods (now known to be fully terrestrial animals (probably tree browsers) rather than the inhabitants of swamps that they were once portrayed as) could not have sustained high metabolic rates without overheating. This objection cannot apply to the bulk of carnivorous and herbivorous dinosaurs which had body weights within the range of living birds and mammals, and there are in any case several troublesome aspects implicit in the idea of ectothermic sauropods.

First, the sauropods were distributed into latitudes which imply some degree of seasonality. Inertial homoiothermy might keep a brontosaur warm for several days, even a week or two, but it is difficult to see how such an animal could avoid cooling to torpid levels during prolonged bad weather, particularly when exposed to heavy rain. Water has a high heat capacity and will strip heat from a warm body. Second, there is a general tendency for weight specific metabolic rate to decrease with increasing body size (Chapter 1), so the energetic cost of endothermy to a sauropod would be proportionately much less than for an elephant or rhino. Sauropods were so large that they might well have had to solve overheating problems anyway, whether they were ectothermic or endothermic. Perhaps they had richly vascularized skins, and well developed behavioural responses to avoid exposure to too much sun. The convergence of ectothermic and endothermic metabolic rates in very large animals also nullifies the suggestion that sauropods could not eat enough to sustain a high enough metabolic rate to be endothermic. Third, newborn sauropods did not weigh 20–30 tons! Little is known about the reproduction and growth of brontosaurs, but it is doubtful whether they were egg layers (any egg would have approached the size at which a shell needs to be so thick to support the egg that it becomes too thick for the youngster to break out); some form of viviparity probably operated. Young sauropods must also have been fairly large, which, taken with the likelihood of viviparity, implies a low level of fecundity (living large reptiles such as turtles, giant tortoises and crocodilians are egg layers, have high fecundity, vulnerable offspring and slow growth). Species which produce few offspring cannot 'afford' to have them grow slowly as this exposes the young and vulnerable animals for too long to predation and disease. It is difficult to see how a herbivore, which would inevitably have a lower absorption efficiency than carnivores such as alligators, could grow rapidly to 20 tons without being an active forager; high levels of activity at intermediate sizes (say up to 2–3 tons) would demand the high metabolic rate of an endotherm.

Alligators, Komodo dragons, sea turtles and giant tortoises all take several decades to reach maximum size (e.g. Jacobsen and Kushlan, 1989); endotherms grow perhaps 10 times as fast (Case, 1978). As de Ricqle (1974) pointed out, if we are to accept the giant sauropods (up to 15 times the weight of the largest living reptiles) as huge ectotherms, we must postulate great longevity for them, perhaps exceeding 200 years, to allow time for slow, ectothermic growth to the size of sexual maturity. Such longevity is unknown in living animals and it seems far more realistic to accept the arguments of Bakker and de Ricqle, recognizing that endothermy evolved in reptiles, probably several times, in response to the challenge of diurnal and seasonal exposure to low temperature.

The author believes that endothermy originally evolved several times in quite large reptiles, perhaps 20–50 kg in weight. These were exposed to cool conditions at night and warm temperatures during the day. They not only had control of the peripheral circulation (to restrict heat loss) like several living iguanid lizards, but also featured large and powerful leg muscles which functioned aerobically, permitting both long distance steady walking and muscular thermogenesis (like some living varanid monitor lizards — see Bakker, 1971). They probably had some form of insulation, but there is no reason to postulate the evolution of fur at this stage, since animals of this size would maintain reasonable temperature control at night, given a superficial layer of fat (which may have evolved initially as an energy store). There is also no good reason to assume that these early endotherms maintained as high and stable body temperatures as living birds and mammals — a smaller temperature differential between core and the environment would be less expensive in metabolic terms, as would a degree of torpor at night. Coevolution of better control of heat loss/ gain, better limb posture for fast movement, enhanced metabolic rate and heightened body temperature would soon produce an effective homoiothermic endotherm. Given such an animal as a basis, there is no difficulty in deriving the endothermic dinosaurs postulated by Bakker. However, there is still the problem of the evolution of fur and feathers in mammals, pterosaurs and birds. In all three groups, the earliest known representatives were small animals. Early pterosaurs (e.g. *Pterodactylus*) and the toothed Jurassic bird *Archaeopteryx* were no bigger than jackdaws; early mammals, believed to have evolved from the Therapsid or 'mammal-like' reptiles of the late Permian, were even smaller — perhaps comparable in size to oppossums or tree shrews. Given the problems of being endothermic and small (disadvantageous surface-area to volume ratio; difficulty of carrying fat for insulation), it seems likely that insulation derived from modified scales allowed the evolution of small endotherms from larger, non-furry/feathered endotherms. Early birds and pterosaurs needed to be small to fly at all; early mammals seem to have been specialists which occupied niches denied to larger endotherms (e.g. arboreal, fossorial, nocturnal).

9.3 COLD AND EXTINCTION

Cold kills animals. Severe winters in temperate and polar areas have profound effects on populations of birds and mammals, killing many adults and reducing breeding success in survivors. These events may be due to the direct effects of low temperatures, or to indirect influences such as freezing of sources of drinking water or the covering of vegetation by snow. Even short periods of cold can wreak havoc with populations of wrens, titmice and finches in Britain, while caribou starve in Canada if snow covers the lichens too deeply for too long. Decreased populations of herbivorous birds and mammals have consequences for predators too; owl numbers decline as small mammals disappear, while foxes, wolverines and wolves starve to death in bad winters. Unusually low temperatures also affect ectotherms. The severe winter in Britain during 1962–63 produced particularly low coastal temperatures. Crisp and others (1964) reported great reductions in the populations of coastal fish and intertidal animals such as shore crabs, barnacles, cockles, oysters, mussels and limpets, partly because many animals were exposed to temperatures below their lower lethal limits, but also because more hardy species (e.g. the mussel, *Mytilus edulis*) were rendered too torpid to protect themselves against attacks by predators. In both ectotherms and endotherms the effects of a single severe winter may take several years to reverse.

Because cold can have such devastating effects it has often been invoked as the likely cause of past episodes of mass extinction. The disappearance of dinosaurs, pterosaurs, plesiosaurs, mosasaurs and icthyosaurs at the end of the Cretaceous is the most spectacular example of such episodes, but relatively sudden mass extinctions have happened often, the most recent being that of large terrestrial mammals (including mammoths, mastodons, giant sloths and sabre-toothed cats) a few tens or hundreds of thousands of years ago. In most cases the upheavals have not been limited to the populations of large terrestrial animals; the marine ecosystem has been equally affected, particularly in terms of plankton composition and in the demise of large predators.

Explanations for mass extinctions have been legion, but a recurrent theme in the past few decades has been that of climatic change (earlier notions of species-senescence now attract little support). The disappearance of most large reptiles at the end of the Cretaceous (followed by the adaptive radiation of mammals and birds) has attracted the greatest speculation. Many scientists believe that low temperatures paid an important role, changing the vegetation and killing the large terrestrial animals, locking up water in the ice caps (as well as reducing sea volume simply by increasing its density) and thereby draining the warm, productive, shallow coastal seas (thus exterminating the large marine reptiles) as well as creating land bridges to expose populations to novel competition from species not encountered previously. However, there is much dispute about the suddenness of extinctions (did they occur over tens, hundreds, thousands or even millions of years?) and the source of temperature changes. At present there is considerable disagreement between those of a uniformitarian bent (who would accept fairly slow climatic changes resulting from changing patterns of distance from and orientation to the sun) and modern catastrophists such as Walter Alvarez (e.g. Alvarez *et al.*, 1982) who invoke collisions with meteorites or close encounters with asteroids to explain sudden world cooling as vast quantities of steam and dust are thrown into the atmosphere, blocking out much solar heat and illumination, perhaps for several years.

References

Ahlgren, J.A., Cheng, C.C., Schrag, J.D. and Devries, A. (1988) Freezing avoidance and the distribution of antifreeze glycopeptides in body fluids and tissues of antarctic fish. *Journal of Experimental Biology*, **137**, 549–63.

Aitchison, C.W. (1984) A possible subnivean food chain. In *Winter ecology of small mammals*. (ed. J.F. Merritt). Carnegie Museum of Natural History, Pittsburgh, Pennsylvania.

Allen, J.A. (1877) The influence of physical conditions on the genesis of species. *Radical Review*, **1**, 108–40.

Allendorf, F.W. and Thorgaard, G.H. (1984) Tetraploidy and the evolution of salmonid fishes. In *Evolutionary genetics of fish* (ed. B.J. Turner) Plenum Press, New York.

Alvarez, W., Asaro, F., Michel, H.V. and Alvarez, L. (1982) Evidence for a major meteorite impact on the earth 34 million years ago: implications for Eocene extinctions. *Science*, **216**, 886–8.

Arnold, W. (1988) Social thermoregulation during hibernation in alpine marmots. *Journal of Comparative Physiology B*, **158**, 151–6.

Asahina, E. (1969) Frost resistance in insects. *Advances in Insect Physiology*, **6**, 1–49.

Aschoff, J. (1982) The circadian rhythm of body temperature as a function of body size. In *A companion to animal physiology* (eds C.R. Taylor, K. Johansen and L. Bolis) Cambridge University Press, Cambridge.

ASHRAE (1981) *ASHRAE handbook of fundamentals*, Atlanta, Georgia. American Society of Heating, Refrigerating and Air Conditioning Engineers.

Astaurov, B.L. (1969) Experimental polyploidy in animals. *Annual Review of Genetics*, **3**, 99–126.

Augee, M.L., Raison, J.K. and Hulbert, A.J. (1979) Seasonal changes in membrane lipid transitions and thyroid function in the hedgehog. *American Journal of Physiology*, **236**, E589–93.

Aulie, A. and Tøien, Ø. (1988) Threshold for shivering in aerobic and anaerobic muscles in bantam cocks and incubating hens, *Journal of Comparative Physiology B.*, **158**, 431–5.

Avery, R.A. (1979) *Lizards — a study in thermoregulation*. Edward Arnold, London.

Bale, J.S. (1989) Cold hardiness and overwintering of insects. *Agricultural Zoology Reviews*, **3**, 157–2.

Bakker, R.T. (1971) Dinosaur physiology and the origin of mammals. *Evolution*, **25**, 636–58.

Bakker, R.T. (1975a) Dinosaur renaissance. *Scientific American*, **232**, 58–78.

Bakker, R.T. (1975b) Experimental and fossil evidence for the evolution of tetrapod bioenergetics. In *Perspectives of biophysical ecology* (eds D.M. Gates and R.B. Schmerl). Springer-Verlag, New York.

Bakker, R.T. (1986) *The dinosaur heresies: a revolutionary view of dinosaurs.* Longman Scientific and Technical, Harlow, Essex.

Bakko, E.B., Porter, W.P. and Wunder, B.A. (1988) Body temperature patterns in black-tailed prairie dogs in the field. *Canadian Journal of Zoology*, **66**, 1783–9.

Barnes, B.M. (1989) Freezing avoidance in a mammal: body temperatures below 0°C in an arctic hibernator. *Science*, **244**, 1593–5.

Bartholomew, G.A. (1966) A field study of the temperature relations in the Galapagos marine iguana. *Copeia*, **1966**, 241–50.

Bartholomew, G.A. and Lasiewski, R.C. (1965) Heating and cooling rates, heart rate and simulated diving in the Galapagos marine iguana. *Comparative Biochemistry and Physiology*, **16**, 573–82.

Bailey, S.E.R. and Lazaridou-Dimitriadou, M. (In press). Inverse temperature acclimation of heart rate in hibernating land snails. *Journal of Molluscan Studies.*

Baust, J.G. and Morrissey, R.E. (1975) Supercooling phenomenon and water content independence in the overwintering beetle, *Coleomegilla maculata. Journal of Insect Physiology*, **21**, 1751–4.

Baust, J.G. and Rojas, R.R. (1985) Review — insect cold hardiness: facts and fancy. *Journal of Insect Physiology*, **31**, 755–9.

Baust, J.G., Rojas, R.R. and Hamilton, M.D. (1985) Life at low temperatures: representative insect adaptations. *Cryo-letters*, **6**, 199–210.

Bayne, B.L. (1976) *Marine mussels: their ecology and physiology.* Cambridge University Press, Cambridge.

Belyaev, D.K. and Borodin, P.M. (1982) The influence of stress on variation and its role in evolution. *Biologi Zentralbl*, **100**, 705–14.

Bennet, A.F. and Dawson, W.R. (1976) Metabolism. In *Biology of the reptilia* Vol. 5 (eds C. Gans and W.R. Dawson). Academic Press, New York.

Bergmann, C. (1847). Uber die verhaltniesse der warmeokonomie der thier zu ihrer grosse. *Gottingen Studies*, **1**, 595–708.

Berk, M.L. and Heath, J.E. (1975) An analysis of behavioural thermoregulation in the lizard, *Dipsosaurus dorsalis. Journal of Thermal Biology*, **1**, 15–22.

Block, W. (1982) Cold hardiness in invertebrate poikilotherms. *Comparative Biochemistry and Physiology*, **73A**, 581–93.

Block, W. and Young, S.R. (1979) Measurement of supercooling in small arthropods and water droplets. *Cryo-letters*, **1**, 85–91.

Bogart, J.P. (1980) Polyploidy in evolution of amphibians and reptiles. In *Polyploidy: Biological Relevance* (ed. H.L. Lewis). Plenum Press, New York.

Bogert, C.M. (1953) Body temperature of the tuatara under natural conditions. *Zoologica*, **38**, 63–4.

Boggs, D.F., Kilgore, D.L. Jr and Birchard, G.F. (1984) Respiratory physiology of burrowing mammals and birds. *Comparative Biochemistry and Physiology*, **77A**, 1–7.

Brafield, A.E. and Llewellyn, M.J. (1982) *Animal energetics*. Blackie & Son Ltd, Glasgow.

Brandt, D. Ch. (1980) Is the mound of *Formica polyctena* Foerst, in origin a simulation of a rock? *Oecologia*, **44**, 281–2.

Brattstrom, B.H. (1979) Amphibian temperature regulation studies in the field and in the laboratory. *American Zoologist*, **19**, 345–56.

Brian, M.V. (1977) *Ants*. Collins, London.

Brody, A.J. and Pelton, M.R. (1988) Seasonal changes in digestion in black bears. *Canadian Journal of Zoology*, **66**, 1482–4.

Burlington, R.F., Bowers, W.D., Daum, R.C. and Ashbaugh, P. (1972) Ultrastructure changes in heart tissue during hibernation. *Cryobiology*, **9**, 224–8.

Burlington, R.F. and Darvish, A. (1988) Low-temperature performance of isolated working hearts from a hibernator and a nonhibernator. *Physiological Zoology*, **61**, 387–95.

Burlington, R.F. and Shug, A.L. (1981) Seasonal variation of carnitine levels of the ground squirrel *Citellus tridecemlineatus*. *Comparative Biochemistry and Physiology*, **68B**, 431–6.

Burton, A.C. and Edholm, O.G. (1955) *Man in a cold environment: physiological and pathological effects of exposure to low temperatures*. Edward Arnold, London.

Butler, C.G. (1974) *The world of the honey bee*. (2nd edn). Collins, London.

Cannon, R.J.C. and Block, W. (1988) Cold tolerance of microarthropods. *Biological Reviews*, **63**, 23–77.

Carey, F.G. and Teal, J.M. (1966) Heat conservation in tuna fish muscle. *Proceedings of the National Academy of Sciences of the U.S.A.*, **56**, 1464–9.

Carpenter, F.L. (1974) Torpor in an Andean hummingbird. Its ecological significance. *Science*, **183**, 545–7.

Case, T.J. (1978) On the evolution and adaptive significance of postnatal growth rates in the terrestrial vertebrates. *Quarterly Review of Biology*, **53**, 243–82.

Casey, T.M. (1981) Nest insulation: energy savings to brown lemmings using a winter nest. *Oecologia*, **50**, 199–204.

Charland, B. (1987) An examination of factors influencing first year recruitment in the northern Pacific rattlesnake, *Crotalus viridis oreganus*, in British Columbia. MSc Thesis: University of Victoria, Victoria BC.

Childress, J.J. and Somero, G.N. (1979) Depth-related enzymic activities in muscle, brain, and heart of deep living pelagic marine teleosts. *Marine Biology*, **52**, 273–83.

Clarke, A. (1980) A reappraisal of the concept of metabolic cold adaptation in polar marine invertebrates. *Biological Journal of the Linnean Society*, **14**, 77–92.

Clarke, A. (1983) Life in cold water: the physiological ecology of polar marine ectotherms. *Oceanography and Marine Biology Annual Review*, **21**, 341–453.

Cloudsley-Thompson, J.L. (1970) *The temperature and water relations of reptiles*. Merrow, Watford, Hertfordshire, England.

Cogger, H.G. and Holmes, A. (1960) Thermoregulatory behaviour in a specimen of *Morelia spilotes variegata* Gray (Serpentes: Boidae). *Proceedings of the Linnean Society of New South Wales*, **85**, 328–33.

Collins, B.G. and Rebelo, T. (1987) Pollination biology of the *Proteaceae* in Australia and southern Africa. *Australian Journal of Ecology*, **12**, 387–421.

Collins, B.G., Wooller, R.D. and Richardson, K.C. (1987) Torpor by the honey possum, *Tarsipes rostratus* (Marsupialia: Tarsipedidae), in response to food shortage and low environmental temperature. *Australian Mammalogy*, **11**, 51–7.

Constanzo, J.P. (1988) Effects of humidity, temperature, and submergence behaviour of survivorship and energy use in hibernating garter snakes, *Thamnophis sirtalis*. *Canadian Journal of Zoology*, **67**, 2486–92.

Constanzo, J.P. (1989) A physiological basis for prolonged submergence in hibernating garter snakes *Thamnophis sirtalis*: evidence for an energy-sparing adaptation. *Physiological Zoology*, **62**, 580–92.

Contreras, L.C. (1984) Bioenergetics of huddling: test of a psycho-physiological hypothesis. *Journal of Mammalogy*, **65**, 256–62.

Coon, C.S., Garn, S.M. and Birdsell, J.B. (1950) *Races*. Charles C. Thomas, Springfield, Illinois.

Coope, G.R. (1969) The contribution that coleoptera of glacial Britain could have made to the subsequent colonization of Scandinavia. *Opuscula entomologica (Lund)*, **34**, 95–108.

Cossins, A.E. and Sinensky, M. (1984) Adaptations of membranes to temperature, pressure and exogenous lipids. In *Physiology of Membrane fluidity*. Vol. 1 (ed. M. Shinitsky). CRC Press Inc., Boca Raton, Florida.

Cott, H.B. (1961) Scientific results of an inquiry into the ecology and economic status of the Nile crocodile (*Crocodilus niloticus*) in Uganda and northern Rhodesia. *Transactions of the Zoological Society of London*, **29**, 211–356.

Cowles, R.B. and Bogert, C.M. (1947) Temperature of desert reptiles. *Science*, **105**, 282.

Crisp, D.J. *et al.* (1964) The effects of the severe winter of 1962–63 on marine life in Britain (ed. D.J. Crisp). *Journal of Animal Ecology*, **33**, 165–210.

Crisp, D.J., Davenport, J. and Gabbott, P.A. (1977) Freezing tolerance in *Balanus balanoides*. *Comparative Biochemistry and Physiology*, **57A**, 359–61.

Dalenius, P. (1965) The acarology of the Antarctic regions. *Monographiae biologica*, **14**, 414–30.

Davenport, J. (1977) A study of the effects of copper applied continuously and discontinuously to specimens of *Mytilus edulis* (L.) exposed to steady and fluctuating salinity levels. *Journal of the Marine Biological Association of the United Kingdom*, **57**, 907–24.

Davenport, J. (1979a) The isolation response of mussels, *Mytilus edulis* (L.), exposed to falling seawater concentrations. *Journal of the Marine Biological Association of the United Kingdom*, **59**, 123–32.

Davenport, J. (1979b) Cold resistance in *Gammarus duebeni* Liljeborg. *Astarte*, **12**, 21–6.

Davenport, J. (1981) The opening response of the mussel (*Mytilus edulis* (L.)) exposed to rising seawater concentrations. *Journal of the Marine Biological Association of the United Kingdom*, **61**, 667–78.

Davenport, J. (1982) Environmental simulation experiments on marine and estuarine animals. *Advances in Marine Biology*, **19**, 133–256.

Davenport, J. (1985a) Osmotic control in marine animals. In *Physiological adaptations of marine animals* (ed. M.S. Laverack). *Symposium of the Society for Experimental*

Biology Number XXXIX. pp. 207–44. The Company of Biologists, Cambridge.

Davenport, J. (1985b) *Environmental stress and behavioural adaptation.* Croom Helm, London.

Davenport, J. (1988) Oxygen consumption and ventilation rate at low temperatures in the antarctic protobranch bivalve mollusc *Yoldia* (=*Aequiyoldia*) *eightsi* (Courthouy). *Comparative Biochemistry and Physiology*, **90A**, 511–3.

Davenport, J. (1989) The effects of salinity and low temperature on eggs of the Icelandic capelin *Mallotus villosus. Journal of the Marine Biological Association of the United Kingdom*, **69**, 1–9.

Davenport, J. and Carrion-Cotrina, M. (1981) Responses of the mussel, *Mytilus edulis* (L.) to simulated subarctic tidepool conditions. *Journal of Thermal Biology*, **6**, 257–65.

Davenport, J., Grove, D.J., Cannon, J. Ellis, T.R. and Stables, R. (1990a) Food capture, appetite, digestion rate and efficiency in hatchling and juvenile *Crocodylus porosus* Schneider. *Journal of Zoology*, **220**, 569–92.

Davenport, J., Holland, D.L. and East, J. (1990b) Thermal and biochemical characteristics of the lipids of the leatherback turtle *Dermochelys coriacea* (L.): evidence of endothermy. *Journal of the Marine Biological Association of the United Kingdom*, **70**, 33–41.

Davenport, J., Lønning, S. and Kjørsvik, E. (1981) Osmotic and structural changes during early development of eggs and larvae of the cod, *Gadus morhua* L. *Journal of Fish Biology*, **19**, 317–31.

Davenport, J., Lønning, S. and Kjørsvik, E. (1986) Some mechanical and morphological properties of the chorions of marine teleost eggs. *Journal of Fish Biology*, **29**, 289–301.

Davenport, J. and Sayer, M.D.J. (1989) Observations on the aquatic locomotion of young salt-water crocodiles (*Crocodilus porosus* Schneider). *Herpetological Journal*, **1**(8), 356–61.

Davenport, J. and Stene, A. (1986) Freezing resistance, temperature and salinity tolerance in eggs, larvae and adults of capelin, *Mallotus villosus*, from Balsfjord. *Journal of the Marine Biological Association of the United Kingdom*, **66**, 145–57.

Davenport, J. and Vahl, O. (1983) Desiccation resistance in the eggs of the capelin *Mallotus villosus. Astarte*, **12**, 35–7.

Davenport, J., Vahl, O. and Lønning, S. (1979) Cold resistance in the eggs of the capelin *Mallotus villosus. Journal of the Marine Biological Association of the United Kingdom*, **59**, 443–53.

Davenport, J. and Woolmington, A. (1981) Behavioural responses of some rocky shore fish exposed to adverse environmental conditions. *Marine Behaviour and Physiology*, **8**, 1–12.

Davis, W.H. (1970) Hibernation: ecology and physiological ecology. In *Biology of bats* (ed. W.A. Wimsatt). vol 1, Academic Press, London.

Davis, W.H. and Reite, O.B. (1967) Responses of bats from temperate regions to changes in ambient temperatures. *Biological Bulletin*, **132**, 320–8.

Dawson, T.J. (1983) *Monotremes and marsupials: the other mammals.* Edward Arnold, London.

Dawson, W.R. and Hudson, J.W. (1970) Birds. In *Comparative physiology of thermoregulation* Vol. 1 (ed. G. Causey Whittow) Academic Press, London.

De Freitas, C.R. and Symon, L.V. (1987) A bioclimatic index of human survival times in the Antarctic. *Polar Record*, **23**(147), 651–9.

Devries, A.L. (1974) Survival at freezing temperatures. In *Biochemical and biophysical perspectives in marine biology* (eds J.S. Sargent and D.W. Mallins). Vol. I. Academic Press, London. 289–330.

Devries, A.L. (1976) Antifreezes in cold-water fishes. *Oceanus*, **19**, 23–31.

Devries, A.L. (1980) Biological antifreezes and survival in freezing environments. In *Animals and environmental fitness* (ed. R. Gilles). Volume 1 (Invited Lectures). Pergamon Press, Oxford.

Devries, A.L. (1982) Antifreeze agents in coldwater fishes. *Comparative Biochemistry and Physiology*, **73A**, 627–40.

Devries, A.L. and Eastman, J.I. (1982) Physiology and ecology of notothenioid fish of the Ross Sea. *Journal of the Royal Society of New Zealand*, **11**, 329–40.

Devries, A.L. and Lin, Y. (1977a) The role of glycoprotein antifreezes in the survival of antarctic fishes. In *Adaptations within antarctic ecosystems*. (ed. G.A. Llano). pp. 439–58. Gulf Publishing Company, Texas, Houston.

Devries, A.L. and Lin, Y. (1977b) Structure of a peptide antifreeze and mechanism of adsorption to ice. *Biochemica et Biophysica Acta*, **495**, 388–92.

Devries, A.L. and Wohlschlag (1969) Freezing resistance in some antarctic fishes. *Science*, **163**, 1074–5.

Dobbs, G.H. and Devries, A.L. (1975a) Renal function in antarctic teleosts fishes: serum and urine composition. *Marine Biology*, **29**, 59–70.

Dobbs, G.H. and Devries, A.L. (1975b) Aglomerular nephron of Antarctic teleosts: a light and electron microscopic study. *Tissue and Cell*, **7**(10) 159–70.

Duman, J.G. (1977a) The role of macromolecular antifreeze in the darkling beetle. *Meracantha contracta. Journal of Comparative Physiology*, **115**, 279–86.

Duman, J.G. (1977b) Variations in macromolecular antifreeze levels in larvae of the darkling beetle, *Meracantha contracta. Journal of Experimental Zoology*, **201**, 85–92.

Duman, J.G. (1979) Thermal hysteresis factors in overwintering insects. *Journal of Insect Physiology*, **25**, 805–10.

Duman, J.G. (1984) Changes in the overwintering mechanism in the Cucujus beetle *Cucujus claviceps. Journal of Insect Physiology*, **30**, 235–9.

Duman, J.G. and Devries, A.L. (1974) The effects of temperature and photoperiod on the production of antifreeze in cold water fishes. *Journal of Experimental Zoology*, **190**, 89–97.

Duman, J.G. and Devries, A.L. (1976) Isolation, characterization and physical properties of protein antifreezes from the winter flounder, (*Pseudopleuronectes americanus*). *Comparative Biochemistry and Physiology*, **53B**, 375–80.

Duman, J.G. and Horwath, K.L. (1983) The role of haemolymph proteins in the cold tolerance of insects. *Annual Review of Physiology*, **45**, 261–70.

Duman, J.G., Horwarth, K.L., Tomchaney, A. and Patterson, J.L. (1982) Antifreeze agents of terrestrial arthropods. *Comparative Biochemistry and Physiology*, **73A**, 545–55.

Dunn, J.F. (1988) Low-temperature adaptation of oxidative energy production in cold-water fishes. *Canadian Journal of Zoology*, **66**, 1098–104.

Edholm, O.G. (1978) *Man — hot and cold*. Edward Arnold, London.

Edholm, O.G. and Weiner, J.S. (1981) *The principles and practice of human physiology*.

Academic Press, London.

Eisentraut, M. (1934) Der Winterschlaf der Fledermäuse mit bersonderer Berücksichtigung der Wärmeregulation. *Z. Morphol. Oekol. Tiere*, **29**, 231–67.

Eldredge, N. and Gould, S.J. (1972) Punctuated equilibria: an alternative to phyletic gradualism. In *Models in palaeobiology* (ed. T.J.M. Schopf). Freeman, Cooper & Co., San Francisco.

Emiliani, C. (1972) Quaternary palaeotemperatures and the duration of high temperature intervals. *Science*, **178**, 398–401.

Fletcher, G.L. (1979) The effects of hypophysectomy and pituitary replacement on the plasma freezing point depression, Cl^-, glucose and protein anti-freeze in the winter flounder (*Pseudopleuronectes americanus*). *Comparative Biochemistry and Physiology*, **63A**, 535–7.

Fletcher, N.H. (1970) *The chemical physics of ice*. Cambridge University Press, Cambridge.

Folk, G.E. (1957) Twenty-four hour rhythms of mammals in a cold environment. *American Naturalist*, **91**, 153–66.

Folk, G.E. (1969) *Introduction to environmental physiology*. Lea and Febinger, New York.

Frair, W., Ackman, R.G. and Mrosovsky, N. (1972) Body temperature of *Dermochelys coriacea*: warm turtle from cold water. *Science*, **177**, 791–3.

Franks, F. (1982) The properties of aqueous solutions at sub-zero temperatures. In *Water — a comprehensive treatise*. Vol. 7 (ed. F. Franks). Plenum Press, New York.

Franks, F. (1985) *Biophysics and biochemistry at low temperatures*. Cambridge University Press, Cambridge.

Freeland, W.J. and Jansen, D.H. (1974) Strategies in herbivory by mammals: the role of plant secondary compounds. *American Naturalist*, **108**, 269–89.

French, N.R. and Hodges, R.W. (1959) Torpidity in cave-roosting hummingbirds. *Condor*, **61**, 223.

Friis-Sørensen (1983) Unpublished Cand. real. thesis, University of Tromsø.

Frisch, J., Øritsland, N.A. and Krog, J. (1974) Insulation of furs in water. *Comparative Biochemistry and Physiology*, **47A**, 403–10.

Geiser, F. (1988a) Reduction of metabolism during hibernation and daily torpor in mammals and birds: temperature effect or physiological inhibition? *Journal of Comparative Physiology B*, **158**, 25–37.

Geiser, F. (1988b) Daily torpor and thermoregulation in *Antechinus* (Marsupialia): influence of body mass, season, development, reproduction and sex. *Oecologia*, 77, 395–9.

Geiser, F. and Kenagy, G.J. (1988) Torpor duration in relation to temperature and metabolism in hibernating ground squirrels. *Physiological Zoology*, **61**, 442–9.

Gilles-Baillen, M. (1979) Hibernation and membrane function in reptiles. In *Animals and environmental fitness*. (ed. R. Gilles). Pergamon Press, London.

Glass, M.L., Hicks, J.W., and Riedesel, M.L. (1979) Respiratory responses to long-term temperature exposure in the box turtle, *Terrapene ornata*. *Journal of Comparative Physiology*, **131**, 353–9.

Gloger, C.L. (1833) *Das abandern der vogel durch einfluss des klimas*. Breslau.

Goodenough, U. (1978) *Genetics*. Holt-Saunders, Tokyo.

Gordon, M.S., Amdur, B.H. and Scholander, P.F. (1962) Freezing resistance in some northern fishes. *Biological Bulletin of the Marine Laboratory of Woods Hole, Massachusetts*, **122**, 52–62.

Grav, H.J. and Blix, A.S. (1975) Brown adipose tissue — a factor in the survival of harp seal pups. In *Depressed metabolism and cold thermogenesis*. (ed. L. Jansky). Charles University, Prague.

Grav, H.J., Borsch-Johnsen, B., Dahl, H.A., Gabrielsen, G.W. and Steen, J.B. (1988) Oxidative capacity of tissues contributing to thermogenesis in eider (*Somateria mollissima*) ducklings: changes associated with hatching. *Journal of Comparative Physiology B*, **158**, 513–8.

Greenwald, O.E. (1974) Thermal dependence of striking and prey capture by gopher snakes. *Copeia*, **1974**, 141–8.

Gregory, P.T. (1982) Reptilian hibernation. In *Biology of the reptilia*. Vol. 13 (eds C. Gans and F.H. Pough). Academic Press, New York.

Griggio, M.A. (1982) The participation of shivering and nonshivering thermogenesis in warm and cold-acclimated rats. *Comparative Biochemistry and Physiology*, **73A**, 481–4.

Grimstone, A.V., Mullinger, A.M. and Ramsay, J.A. (1968) Further studies on the rectal complex of the mealworm, *Tenebrio molitor* (Coleoptera, Tenebrionidae). *Philosophical Transactions of the Royal Society Series B*, **253**, 343–82.

Haim, A., Fairall, N. and Prinsloo, P.W. (1985) The ecological significance of calcium bicarbonate in the urine of subterranean rodents: testing a hypothesis. *Comparative Biochemistry and Physiology*, **82A**, 867–9.

Haim, A., Ellison, G.T.H. and Skinner, J.D. (1988) Thermoregulatory circadian rhythms in the pouched mouse (*Saccostomus campestris*). *Comparative Biochemistry and Physiology*, **91A**, 123–7.

Hammel, H.T. (1955) Thermal properties of fur. *American Journal of Physiology*, **182**, 369–76.

Hammel, H.T. (1964) Terrestrial animals in cold; recent studies of primitive man. In *Handbook of Physiology, Adaptations of the Environment* (ed. D.B. Dirrel *et al.*). American Physiological Society, Washington D.C.

Hargens, A.R. and Shabica, S.V. (1973) Protection against lethal freezing temperature by mucus in an Antarctic limpet. *Cryobiology*, **10**, 331–7.

Hardy, R.N. (1976) *Homeostasis*. Edward Arnold, London.

Hasselrot, T.B. (1960) Studies on Swedish bumblebees (Genus *Bombus* Latr.) their domestication and biology. *Opuscula Entomology* (Suppl.) **17**, 1–203.

Havera, S.P. (1979) Temperature variation in a fox squirrel nest box. *Journal of Wildlife Management*, **43**, 252–3.

Hayward, J.S. (1965) Microclimate temperature and its adaptive significance in six geographic races of *Peromyscus*. *Canadian Journal of Zoology*, **43**, 341–50.

Hayward, J.S. and Lyman, C.P. (1967) Non-shivering heat production during arousal from hibernation and evidence for the contribution of brown fat. In *Proceedings III. International symposium on natural mammalian hibernation*. (ed. K.C. Fisher). Oliver and Boyd, Edinburgh.

Hazel, J.R. and Sellner, P.A. (1980) The regulation of membrane lipid composition in thermally-acclimated poikilotherms. In *Animals and environmental fitness. Physiological and biochemical aspects of adaptation and ecology. Vol. 1.* (ed. R. Gilles). Pergamon Press, Oxford.

Heldmaier, G. (1971) Relationship between non-shivering thermogenesis and body size. In *Nonshivering thermogenesis. Proceedings of the symposium.* (ed. L. Jansky). Swets and Zeitlinger N.V, Amsterdam.

Heldmaier, G. and Steinlechner, S. (1981) Seasonal control of energy requirements for thermoregulation in the Djungarian hamster (*Phodopus sungorus*), living in natural photoperiod. *Journal of Comparative Physiology*, **142**, 429–37.

Hemmingsen, A.M. (1960) Energy metabolism as related to body size and respiratory surfaces, and its evolution. *Report. Steno Memorial Hospital*, **9**, 1–110.

Heath, J.E. (1964) Head-body temperature differences in horned lizards. *Physiological Zoology*, **37**, 273–9.

Henriques, V. and Hansen, C. (1901) Vergleischende untersuchungen uber die chemische zusammensetzung des thierischen fettes. *Skandinavian Archives of Physiology*, **11**, 151 *et seq.*

Hill, J.E. and Smith, J.D. (1984) *Bats. A natural history.* British Museum (Natural History), London.

Hille, B. (1984) *Ionic channels of excitable tissues.* Sinauer, Sunderland, Massachusetts, USA.

Hirsch, K.R. and Holzapfel, W.B. (1984) Symmetrical hydrogen bonds in ice-X. *Physics Letters*, **A101**, 142–4.

Hobbs, J. (1974) *Ice physics.* Oxford University Press, Oxford.

Hochachka, P.W. (1988) Channels and pumps — determinants of metabolic cold adaptation strategies. *Comparative Biochemistry and Physiology*, **90B**, 515–9.

Hochachka, P.W. and Somero, G.N. (1973) *Strategies of biochemical adaptation.* W.B. Saunders & Co., New York.

Hochachka, P.W., Emmet, B. and Suarez, R.K. (1988) Limits and constraints in the scaling of oxidative and glycolytic enzymes in homeotherms. *Canadian Journal of Zoology*, **66**, 1128–38.

Hock, R.J. (1951) *Science in Alaska. Proceedings of the second Alaskan Scientific Conference.* AAAS, Alaska Division, Washington DC.

Hock, R.J. (1957) Hibernation. In *Cold injury.* (ed. M.I. Ferrer). Josiah Macy, New York.

Hocking, B. and Sharplin, C.D. (1965) Flower basking by arctic insects. *Nature*, **206**, 215.

Hoesslin, H. von (1888) Uber die ursache der scheinbaren abhangigkeit des umsatzes von der grosse der korperoberflache. *Arch. Anat. Physiol., Physiol. Abth.* p. 323.

Holliday, F.G.T. (1969) The effects of salinity on the eggs and larvae of teleosts. In *Fish Physiology*, Vol 1 (eds W.S. Hoar and D.J. Randall) pp. 293–311. Academic Press, New York.

Holliday, F.G.T. and Jones, M.P. (1967) Some effects of salinity on the developing eggs and larvae of the plaice (*Pleuronectes platessa*). *Journal of the Marine Biological Association of the United Kingdom*, **47**, 39–48.

Hoo-Paris, R., Castex, C.H. and Siutter, R.CH. (1978) Plasma glucose and insulin in the hibernating hedgehog. *Diabetes and Metabolism* (Paris), **4**, 13–18.

Hotton, N. (1980) An alternative to dinosaur endothermy: the happy wanderers. In *A cold look at the warm-blooded dinosaurs.* (eds R.D.K. Thomas, and E.C. Olson). Westview Press Inc., Boulder, Colorado.

Houlihan, D.F. and Allan, D. (1982) Oxygen consumption of some Antarctic and

British gastropods: an evaluation of cold adaptation. *Comparative Biochemistry and Physiology*, **73A**, 383–7.

Hudson, J. (1978) Shallow daily torpor: a thermoregulatory adaptation. In *Strategies in cold: natural torpidity and thermogenesis*. (eds L.C.H. Wang, and J.W. Hudson). Academic Press, New York.

Hudson, J. (1981) Role of the endocrine glands in hibernation with special reference to the thyroid gland. In *Survival in the cold: hibernation and other adaptations*. (eds X.J. Musacchia and L. Jansky). Elsevier, New York.

Huey, R.B. and Slatkin, M. (1976) Costs and benefits of lizard thermoregulation. *Quarterly Review of Biology*, **51**, 363–84.

Hutchinson, V.H., Dowling, H.G. and Vinegar, A. (1966) Thermoregulation in a brooding female Indian python, *Python molurus brivittatus*. *Science*, **151**, 694–6.

Ibing, V.J. and Theede, H. (1975) Zur Gefrierresistenz litoraler Mollusken von der deutschen Nordseekuste. *Kieler Meeresforschungen*, **31**, 44–8.

Innes, S. and Lavigne, D.M. (1979). Comparative energetics of coat colour polymorphisms in the eastern grey squirrel *Sciurus carolinensis*. *Canadian Journal of Zoology*, **57**, 585–92.

Ishay, J., Bytinski-Salz, H. and Shulov, A. (1967) Contributions to the bionomics of the oriental hornet (*Vespa orientalis* Fab.). *Israel Journal of Entomology*, **2**, 45–106.

Irving, S. and Krog, J. (1955) Temperature of skin in the arctic as a regulator of heat. *Journal of Applied Physiology*, **7**, 355–64.

Irving, S., Schmidt-Nielsen, K. and Abrahamsen, N.S.B. (1957) On the melting points of animal fats in cold climates. *Physiological Zoology*, **30**, 93–105.

Ivlev, V.S. (1960) Analysis of the mechanism of distribution of marine fishes under the conditions of a temperature gradient. *Zool. Zh.*, **39**, 494–9.

Jacobsen, T. and Kushlan, J.A. (1989) Growth dynamics of the American alligator (*Alligator mississippiensis*). *Journal of Zoology*, **219**, 309–28.

Jansky, L. (1973) Non-shivering thermogenesis and its thermoregulatory significance. *Biological Reviews*, **48**, 85–132.

Jansky, L., Kahlerova, Z., Nedoma, J. and Andrews, J.F. (1981) Humoral control of hibernation in gold hamsters. In *Survival in the cold: hibernation and other adaptations*. (eds X.J. Musacchia and L. Jansky). Elsevier, New York.

Jeffers (1931) Unpublished PhD thesis, University of Toronto.

Johnson, F.H., Eyring, H. and Stover, B.J. (1974) *The theory of rate processes in biology and medicine*. Wiley, New York.

Johnston, I.A. and Altringham, J.D. (1988) Muscle contraction in polar fishes: experiments with demembranated muscle fibres. *Comparative Biochemistry and Physiology*, **90B**, 547–55.

Jouventin, P. (1975) Mortality parameters in emperor penguins *Aptenodytes*. In *The biology of penguins* (ed. B. Stonehouse). Macmillan Press, London.

Kalbuchov, I.N. (1935) Anabiose bei Wirbeltieren und Insekten bei Temperaturen unter 0°. *Zool. Jahrb., Abt., Allgem. Zool. Physiol. Tiere*, **55**, 47–64.

Kanwisher, J.W. (1955) Freezing in intertidal animals. *Biological Bulletin of Woods Hole, Mass.*, **109**, 56–63.

Kanwisher, J.W. (1959) Histology and metabolism of frozen intertidal animals. *Biological Bulletin of Woods Hole, Mass.*, **116**, 258–64.

Kanwisher, J.W. and Sundnes, G. (1966) Thermal regulations in cetaceans. pp. 398–409.

In *Whales, dolphins and porpoises* (ed. K.S. Norris). University of California Press, Berkley, CA.

Karow, A.M. and Webb, W.R. (1965) Tissue freezing. *Cryobiology*, **2**, 99–108.

Keatinge, W.E. (1969) *Survival in cold water.* Blackwell Scientific Publications, Oxford.

Kent, J., Koban, M. and Prosser, C.L. (1988) Cold acclimation-induced protein hypertrophy in channel catfish and green sunfish. *Journal of Comparative Physiology B*, **158**, 185–98.

Kergis, J.J. (1975) Some problems of spontaneous and induced mutagenesis in mammals and man. *Mutation Research*, **29**, 271–7.

Kjørsvik, E., Davenport, J. and Lønning, S. (1984) Osmotic changes during the development of eggs and larvae of the lumpsucker, *Cyclopterus lumpus* L. *Journal of Fish Biology*, **24**, 311–21.

Kollias, J., Bartlett, L., Bergsteinova, V., Skinner, J.S., Buskirk, E.R. and Nicholas, W.C. (1974) Metabolic and thermal responses of women during cooling in water. *Journal of Applied Physiology*, **36**, 577–80.

Korhonen, H., Harri, M. and Hohtola, E. (1985) Response to cold in the blue fox and racoon dog as evaluated by metabolism, heart rate and muscular shivering: a re-evaluation. *Comparative Biochemistry and Physiology*, **82A**, 959–64.

Krog, J.O., Zachariassen, K.E., Larsen, B. and Smidsrod, O. (1979) Thermal buffering in Afro-Alpine plants due to nucleating agent-induced water freezing. *Nature, London*, **282**, 300–1.

Krogh, A. (1916) *The respiratory exchange of animals and man.* Longmans, London.

Krüger, K., Prinzinger, R. and Schuchmann, K-L. (1982) Torpor and metabolism in humming birds. *Comparative Biochemistry and Physiology*, **73A**, 679–89.

Kruuk, H. and Balharry, D. (1990) Effects of sea water on thermal insulation of the otter, *Lutra lutra. Journal of Zoology, London*, **220**, 405–15.

Kshatriya, M. and Blake, R.W. (1988) Theoretical model of migration energetics in the blue whale *Balaenoptera musculus. Journal of Theoretical Biology*, **133**, 479–98.

Kukal, O. and Duman, J.G. (1989) Switch in the overwintering strategy of two insect species and latitudinal differences in cold hardiness. *Canadian Journal of Zoology*, **67**, 825–7.

Kunz, T.H. (1982) *Ecology of Bats.* Plenum Press, New York.

Kurta, A. (1985) External insulation available to non-nesting mammal, the little brown bat (*Myotis lucifugus*). *Comparative Biochemistry and Physiology*, **82A**, 413–20.

Laidler, K. (1980) *Squirrels in Britain.* David and Charles, Newton Abbot.

Leakey, R.E. (1981) *The making of mankind.* Dutton, New York.

Lee, R.E. Jr., Zachariassen, K.E. and Baust, J.G. (1981) Effect of cryoprotectants on the activity of haemolymph nucleating agents in physical solutions. *Cryobiology*, **18**, 511–4.

Lehn-Jensen, H. (1986) *Cryopreservation of bovine embryos.* A S Carl Fr. Mortensen, Copenhagen.

Lin, Y. (1979) Environmental regulation of gene expression. *Journal of Biological Chemistry*, **254**, 1422–6.

Lin, Y. and Long, D.J. (1980) Purification and characterization of winter flounder antifreeze peptide messenger ribonucleic acid. *Biochemistry*, **19**, 1111–6.

Lin, Y., Raymond, J.A., Duman, J.G. and Devries, A.L. (1976) Compartmentalization of NaCl in frozen solutions of antifreeze glycoproteins. *Cryobiology*, **13**, 334–40.

Lincoln, R.F. (1981) Sexual maturation in female triploid plaice, *Pleuronectes platessa*, and plaice x flounder, *Platichthys flesus*, hybrids. *Journal of Fish Biology*, **19**, 497–507.

Lindgren, D. (1972) The temperature influence on the spontaneous mutation rate. I. Literature review. *Hereditas*. **70**, 165–78.

Lindow, S.E. (1983) The role of bacterial ice nucleation in frost injury to plants. *Annual Review of Phytopathology*, **21**, 363–84.

Lindroth, C.H. (1969) The theory of glacial refugia in Scandinavia. Comments on present opinions. *Notulae entomologica (Helsinki)*, **49**, 178–92.

Lindroth, C.H. (1972) Reflections on glacial refugia. *Ambio Special Report*, **2**, 51–4.

Lockyer, C. (1981) Growth and energy budgets of large baleen whales from the southern hemisphere. *FAO Fisheries Series No. 5*, **3**, 379–487.

Lockyer, C.H., McConnell, L.C. and Waters, T.D. (1984) The biochemical composition of fin whale blubber. *Canadian Journal of Zoology*, **62**, 2553–62.

Lønning, S. and Davenport, J. (1980) The swelling egg of the long rough dab, *Hippoglossoides limandoides limandoides* (Bloch). *Journal of Fish Biology*, **17**, 359–78.

Lovelock, J.E. (1953) The haemolysis of human red blood cells by freezing and thawing. *Biochemica et biophysica Acta*, **10**, 414–26.

Lyman, C.P. (1970) Thermoregulation and metabolism in bats. In *Biology of bats*. Vol. 1 (ed. W.A. Wimsatt). Academic Press, New York.

Lyman, C.P. (1982) Mechanisms of arousal. In *Hibernation and torpor in mammals and birds*. (eds C.P. Lyman, J.S. Willis, A. Malan, and L.C.H. Wang). Academic Press, New York.

Lyman, C.P. and Chatfield, P.O. (1955) Physiology of hibernation in mammals. *Physiological Reviews*, **35**, 403–25.

Lyman, C.P., Willis, J.S., Malan, A. and Wang, L.C.H. (1982) *Hibernation in mammals and birds*. Academic Press, New York.

Mackenzie, A.P. (1977) Non-equilibrium freezing behaviour of aqueous systems. *Philosophical Transactions of the Royal Society, Series B.*, **278**, 167–89.

Mackintosh, N.A. (1966) The distribution of southern blue and fin whales. In *Whales, dolphins and porpoises*, (ed. K.S. Norris) pp. 125–44. University of California Press, Berkeley and Los Angeles.

McNab, B.K. (1978) The evolution of endothermy in the phylogeny of mammals. *American Naturalist*, **112**, 1–21.

McNab, B.K. (1980) On estimating thermal conductance in endotherms. *Physiological Zoology*, **53**, 145–56.

McNab, B.K. and Auffenberg, W. (1976) The effect of large body size on the temperature regulation of the Komodo dragon, *Varanus komodoensis*. *Comparative Biochemistry and Physiology*, **55A**, 345–50.

Madison, D.M. (1984) Group nesting and its ecological and evolutionary significance in overwintering microtine rodents. In *Winter ecology of small mammals* (ed. J.F. Merrit). *Carnegie Museum of Natural History Special Publication*, **10**, 267–74.

Malan, A. (1977) Blood acid-base state at a variable temperature. A graphical representation. *Respiratory Physiology*, **31**, 259–75.

Malan, A. (1979) Enzyme regulation, metabolic rate and acid-base state on hibernation. In *Animals and environmental fitness*. (ed. R. Gilles). Pergamon Press, London.

Malan, A. (1982) Respiration and acid-base state in hibernation. In *Hibernation and torpor in mammals and birds* (eds C.P. Lyman, J.S. Willis and L.C.H. Wang). Academic Press, New York.

Malan, A. (1986) pH as a control factor in hibernation. In *Living in the cold. Physiological and biochemical adaptations.* (eds H.C. Heller, X.J. Musacchia and L.C.H. Wang). Elsevier, New York.

Malan, A., Arens, H. and Waechter, A. (1973) Pulmonary respiration and acid-base state in hibernating marmots and hamsters. *Respiratory Physiology*, **17**, 45–61.

Malan, A., Mioskowski, E. and Calgari, C. (1988) Time course of blood-acid state during arousal from hibernation in the European hamster. *Journal of Comparative Physiology B.*, **158**, 495–500.

Marsigny, B., Bouvier, G. and Foray, J. (1988) Les accidents liés au froid en montagne. *Science and Sports*, **3**, 129–36.

Mason, B.J. (1971) *The physics of clouds.* Clarendon, Press, Oxford.

Mather, K. (1939) Crossing over and heterochromatin in the X-chromosome of *Drosophila melanogaster. Genetics*, **24**, 413–35.

Matthews, L.H. (1978) *The natural history of whales.* Wiedenfeld and Nicolson, London.

Mayer, A. and Nichita, (1929) Sur les variations du metabolisme du lapin et modifcation apres exposition au froid. Variation saisonniere du metabolisme du lapin et modification de la fourrure. *Annales Physiologique, Physicochemique et Biologique*, **5**, 621 *et seq.*

Mayr, E. (1956) Geographical character gradients and climatic adaptability. *Evolution*, **10**, 105–8.

Mayr, E. (1963) *Animal species and evolution.* Harvard University Press, Cambridge, Massachusetts.

Mazur, P. (1970) Cryobiology: the freezing of biological sytems. *Science*, **168**, 939–49.

Mazur, P. (1984) Freezing of living cells: mechanisms and indications. *American Journal of Physiology*, **247**, C125–42.

Meer, W. van der (1984) Physical aspects of membrane fluidity. In *Physiology of Membrane fluidity.* Vol. 1 (ed. M. Shinitsky). CRC Press Inc., Boca Raton, Florida.

Meryman, H.T. (1970) The exceeding of a minimum tolerable cell volume in hypertonic suspension as a cause of freezing injury. In *The frozen cell.* (eds G. Wolstenholme and M. O'Connor). Churchill, London.

Meryman, H.T. (1971) Osmotic stress as a mechanism of freezing injury. *Cryobiology.* **8**, 489–500.

Michener, C.D. (1974) *The social behaviour of bees.* Harvard University Press, Cambridge, Massachusetts.

Midtgård, U. (1980) Heat loss from the feet of mallards *Anas platyrhynchos* and arteriovenous heat exchange in the rete tibiotarsale. *Ibis*, **122**, 354–9.

Midtgård, U. (1981) The rete tibiotarsale and arteriovenous association in the hindlimb of birds: a comparative morphological study on countercurrent heat exchange systems. *Acta Zoologica*, **62**, 67–87.

Midtgård, U. (1989) A morphometric study of structures important for cold resistance in the arctic Iceland gull compared to herring gulls. *Comparative Biochemistry and Physiology*, **93A**, 399–402.

Miller, K. (1982) Cold hardiness strategies of some adult and immature insects overwintering in interior Alaska. *Comparative Biochemistry and Physiology*, **73A**,

595–604.

Miller, K. and Werner, R. (1980) Supercooling to −60°C: an extreme example of freezing avoidance in northern willow gall insects. *Cryobiology*, **17**, 621–2.

Mislin, H. and Vischer, L. (1942) Zur Biologie der Chiroptera. II. Die Temperatur-regulation der überwinterden *Nyctalus noctula* Schreb. *Verh. Schweiz. Natur-forsch. Ges.*, **122**, 131–3.

Morgareidge, K.R. and White, F.N. (1969) Cutaneous vascular changes during heat-ing and cooling in the Galapagos marine iguana. *Nature*, **223**, 587–91.

Mrosovsky, N. (1971) *Hibernation and the hypothalamus*. Appleton-Century-Crofts, New York.

Mrosovsky, N. (1980) Thermal biology of sea turtles. *American Naturalist*, **20**, 531–47.

Murphy, D.J. (1979) A comparative study of the freezing tolerances of the marine snails *Littorina littorea* (L.) and *Nassarius obsoletus* (Say). *Physiological Zoölogy*, **52**, 219–30.

Murphy, D.J. (1983) Freezing resistance in intertidal invertebrates. *Annual review of Physiology*, **45**, 289–99.

Murray, M.D. (1976) In *Marine insects* (ed. L. Cheng). pp. 79–96. Amsterdam, North Holland.

Musacchia, X.J. and Jansky, L. (1981) *Survival in the cold: hibernation and other adaptations*. Elsevier, New York.

Musacchia, X.J. and Deavers, D.R. (1981) The regulation of carbohydrate metabolism in hibernators. In *Survival in the cold: hibernation and other adaptations*. (eds X.J. Musacchia and L. Jansky). Elsevier, New York.

Myers, B.C. and Eells, H.K. (1968) Thermal aggregation in *Boa constrictor*. *Herpetologica*, **24**, 61–6.

Nagel, A. (1977) Torpor in the European white-toothed shrews. *Experientia*, **33**, 1455–6.

Neill, W.H. (1979) Mechanisms of fish distribution in heterothermal environments. *American Zoologist*, **19**, 305–17.

Nelson, R.A., Wahner, H.W., Jones, J.D., Ellefson, R.D. and Zollman, P.E. (1973) Metabolism of bears before, during, and after winter sleep. *American Journal of Physiology*, **2254**, 491–6.

Niebuhr, E. (1974) Triploidy in man: cytogenetical and clinical aspects. *Humangenetik*, **21**, 103–25.

Nilsson, T. (1983) *The Pleistocene: geology and life in the Quaternary ice age*. D. Reidel Publishing Co., Dordrecht.

Oeltgen, P.R. and Spurrier, W.A. (1981) Characterization of a hibernation induction trigger. In *Survival in the cold: hibernation and other adaptations*. (eds X.J. Musac-chia and L. Jansky). Elsevier, New York.

Ohno, S. (1967) *Sex chromosomes and sex-linked genes*. Springer, Heidelberg.

O'Grady, S.M., Clarke, A. and Devries, A.L. (1982) Characterization of glycogen pro-tein biosynthesis in isolated hepatocytes from *Pagothenia borchgrevinki*. *Journal of Experimental Zoology*, **220**, 179–89.

Oikari, A. (1975) Hydromineral balance in some brackish-water teleosts after thermal acclimation, particularly at temperatures near zero. *Annales Zoologica Fennici*, **12**, 215–29.

Oliphant, L.W. (1983) First observation of brown fat in birds. *Condor*, **85**, 350–4.

Olla, B. (1977) Personal communication reported in Devries (1980).

Øritsland, N.A. (1970) Temperature regulation of the polar bear (*Thalarctos maritimus*). *Comparative Biochemistry and Physiology*, **37**, 225–33.

Osuga, D.T. and Feeney, R.T. (1978) Antifreeze glycoproteins from Arctic fish. *Journal of Biological Chemistry*, **253**, 5338–43.

Packard, G.C., Packard, M.J., McDaniel, P.L. and McDaniel, L.L. (1989) Tolerance of hatchling painted turtles to subzero temperatures. *Canadian Journal of Zoology*, **67**, 828–30.

Palokangas, R. and Hissa, R. (1971) Thermoregulation in young black-headed gulls (*Larus ridbundus* L.) *Comparative Biochemistry and Physiology*, **38A**, 743–50.

Parry, D.A. (1949) The structure of whale blubber and its thermal properties. *Quarterly Journal of Microscopical Science*, **90**, 13–26.

Parsons, P.A. (1987) Evolutionary rates under environmental stress. *Evolutionary Biology*, **21**, 311–47.

Pauktis, G.L., Shuman, R.D. and Janzen, F.J. (1989) Supercooling and freeze tolerance in hatchling painted turtles (*Chrysemys picta*). *Canadian Journal of Zoology*, **67**, 1082–4.

Pearson, R. (1978) *Climate and evolution*. Academic Press, New York.

Pearson, O.P. (1960) The oxygen consumption and bioenergetics of harvest mice. *Physiological Zoology*, **33**, 152–60.

Pearson, O.P. and Bradford, D.F. (1976) Thermoregulation of lizards and toads at high altitude in Peru. *Copeia*, **1976**, 155–70.

Pearson, P.P. (1954) Habits of the lizard *Liolaemus multiformis multiformis* at high altitude in Southern Peru. *Copeia* **1954**, 111–16.

Petzel, D. and Devries, A.L. (1979) Renal handling of peptide antifreeze in northern fishes. *The Bulletin of the Mount Desert Island Biological Laboratory*, **19**, 17–19.

Pierotti, R. and Pierotti, D. (1983) Costs of evolution of adult pinnipeds. *Evolution*, **37**, 1087–9.

Plough, H.H. (1917) The effect of temperature on crossing over in *Drosophila*. *Journal of Experimental Zoology*, **24**, 147–209.

Potter, N.B. (1965) *Some aspects of the biology of Vespula vulgaris*. Unpublished Ph.D. thesis: University of Bristol.

Precht, H. (1958) Concepts of the temperature adaptation of unchanging reaction systems of cold blooded animals. In *Physiological adaptation*. (ed. C.L. Prosser). American Physiological Society, Washington D.C.

Prevost, Y.A. (1983) Osprey distribution and subspecies taxonomy. In *Biology and management of bald eagles and ospreys*. (ed. D.M. Bird). Harpell Press, Ste Anne de Bellevue, Quebec.

Prinzinger, R., Lüben, I. and Jackel, S. (1986) Vergleichende Unresuchungen zum Energiestoffwechsel bei Kollibris und Nektavölgen. *Journal of Ornithology*, **127**, 303–13.

Prinzinger, R. and Seidle, K. (1986) Experimenteller Nachweis von Torpor bei jungen Mehlschwalben. *Journal of Ornithology*, **127**, 95–6.

Pritchard, P.C.H. (1979) *Encyclopaedia of turtles*. TFH Publications, New Jersey.

Pruitt, W.O. Jr. (1984) Snow and small mammals. In *Winter ecology of small mammals*. (ed. J.F. Merritt). Carnegie Museum of Natural History, Pittsburgh, Pennsylvania.

Pruppacher, H.R. and Klett, J.D. (1978) *Microphysics of clouds and precipitation*. D. Reidel, London.

Pullainen, E. (1973) Winter ecology of the red squirrel (*Sciurus vulgaris* L.) in north-eastern Lapland. *Annales Zoologica Fennici*, **10**, 437–94.

Quinn, P.J. (1985) A lipid-phase separation model of low-temperature damage to biological membranes. *Cryobiology*, **22**, 128–46.

Rakusa-Suszczewski, S. and McWhinnie, M. (1976) Resistance to freezing by Antarctic fauna: supercooling and osmoregulation. *Comparative Biochemistry and Physiology*, **54A**, 291–300.

Ramsay, J.A. (1964) The rectal complex of the mealworm. *Tenebrio molitor* L. (Coleoptera, Tenebrionidae). *Philosophical Transactions of the Royal Society Series B*, **248**, 279–314.

Rankin, J.C. and Davenport, J. (1981) *Animal Osmoregulation*. Blackie and Son, Glasgow.

Raymond, J.A. and Devries, A.L. (1977) Adsorption inhibition as a mechanism in freezing resistance in polar fishes. *Proceedings of the National Academy of Sciences of the U.S.A.*, **74**(6), 2589–93.

Raymond, J.A., Lin, Y. and Devries, A.L. (1975) Glycoproteins and protein antifreeze in two Alaskan fishes. *Journal of Experimental Zoology*, **193**, 125–30.

Raymont, J.E.G. (1967) *Plankton and Productivity in the Oceans*. Pergamon Press, London.

Reftsie, T., Stoss, J. and Donaldson, E. (1982) Production of all female coho salmon (*Onchorhynchus kisutch*) by diploid gynogenesis using irradiated sperm and cold shock. *Aquaculture*, **29**, 67–82.

Ridgeway, S.H. (1972) *Mammals of the sea: biology and medicine*. Charles C. Thomas, Springfield, MA.

Ricqles, A. de. (1974) Evolution of endothermy: histological evidence. *Evolutionary Theory*, **1**, 51–80.

Ring, R.A. (1981) The physiology and biochemistry of cold tolerance in arctic insects. *Journal of Thermal Biology*, **6**, 219–29.

Roberts, D.F. (1973) *Climate and human variability*. Benjamin/Cummings Publishing Company, Menlo Park, California.

Roberts, D.F. (1978) *Climate and human variability*. 2nd edn, Benjamin/Cummings Publishing Company, Menlo Park, California.

Saint Girons, H. (1980) Thermoregulation in reptiles with special reference to the tuatara and its ecophysiology. *Tuatara*, **24**, 59–80.

Salisbury, F.B. (1984) Light conditions and plant growth beneath snow. In *Winter ecology of small mammals*. (ed. J.F. Merritt). Carnegie Museum of Natural History, Pittsburgh, Pennsylvania.

Salt, R.W. (1957) Natural occurrence of glycerol in insects and its relation to their ability to survive freezing. *Canadian Entomology*, **89**, 491–4.

Salt, R.W. (1959) Role of glycerol in the cold hardening of *Bracon cephi* (Gahan). *Canadian Journal of Zoology*, **37**, 59–69.

Salt, R.W. (1961) Principles of insect cold hardiness. *Annual Review of Entomology*, **6**, 55–74.

Salt, R.W. (1966) Factors affecting nucleation in supercooled insects. *Canadian Journal of Zoology*, **44**, 117–33.

Sargent, J.R., Parkes, R.J., Mueller-Harvey and Henderson, R.J. (1987) Lipid biomarkers in marine ecology. In *Microbes in the Sea*. (ed. M.A. Sleigh). Ellis

Horwood Ltd., Chichester.

Satinoff, E., McEwen, G.N. Jr. and Williams, B.A. (1976) Behavioural fever in newborn rabbits. *Science*, **193**, 1139–40.

Schmaranzer, S. and Stabentheiner, A. (1988) Variability of the thermal behaviour of honeybees on a feeding place. *Journal of Comparative Physiology B*, **158**, 135–41.

Schmid, W.D. (1982) Survival of frogs in low temperature. *Science*, **215**, 697–8.

Schmidt-Nielsen, K. and Espeli A. (1941) Das Knockenmark bein Rind und Schwein. *Kungila Norske Videnskabernes Selskaks Forhand linger*, **14**, 17–20.

Schmidt-Nielsen, K. (1946) Melting points of human fats as related to their location in the body. *Acta Physiologica Scandinavia*, **12**, 110–22.

Schneppenheim, R. and Theede, H. (1980) Isolation and characterization of freezing point depressing peptides from larvae of *Tenebrio molitor*. *Comparative Biochemistry and Physiology*, **67B**, 561–8.

Scholander, P.F., Dam, L. Van, Kanwisher, J.W., Hammel, H.J. and Gordon, M.S. (1957) Supercooling and osmoregulation in Arctic fish. *Journal of Cellular and Comparative Physiology*, **49**, 5–24.

Scholander, P.F., Flagg, W., Walters, V. and Irving, L. (1953) Climatic adaptation in arctic and tropical poikilotherms. *Physiological Zoology*, **26**, 67–92.

Scholander, P.F., Hammel, T.H., Hart, J.S., Lemessurier, D.H. and Steen, J. (1958) Cold adaption in Australian aborigines. *Journal of Applied Physiology*, **13**, 211–18.

Scholander, P.F., Hock, R., Walters, V. and Irving, L. (1950c) Adaptation to cold in arctic and tropical mammals and birds in relation to body temperature, insulation, and basal metabolic rate. *Biological Bulletin*, **99**, 259–71.

Scholander, P.F., Hock, R., Walters, V., Johnson, F. and Irving, L. (1950b) Heat regulation in some arctic and tropical mammals and birds. *Biological Bulletin*, **99**, 237–58.

Scholander, P.F., Walters, V., Hock, R. and Irving, L. (1950a) Body insulation of some arctic and tropical mammals and birds. *Biological Bulletin*, **99**, 225–36.

Schrag, J.D., Cheng, C-H. C., Panico, M., Morris, H.R. and Devries, A.L. (1987) Primary and secondary structure of antifreeze peptides from arctic and antarctic zoarcid fishes. *Biochimica et Biophysica Acta*, **915**, 357–70.

Schuchmann, K-L. and Prinzinger, R. (1988) Energy metabolism, nocturnal torpor, and respiration frequency in a green hermit (*Phaethornis guy*). *Journal of Ornithology*, **129**, 469–72.

Schultz, R.J. (1969) Hybridization, unisexuality, and polyploidy in the teleost *Poeciliopsis* (Poeciliidae) and other vertebrates. *American Naturalist*, **108**, 605–19.

Schwaner, T.D. (1989) A field study of thermoregulation in black tiger snakes (*Notothenis ater niger*: Elaphidae) on the Franklin Islands, South Australia. *Herpetologica*, **45**, 393–401.

Shackleton, N.J. (1987) Oxygen isotopes, ice volume and sea level. *Quaternary Science Reviews*, **6**, 183–90.

Sharov, A.G. (1971) New flying reptiles from the Mesozoic of Kazakhstan and Kirghizia. *Trudy Palaeotologicheskogo Instituta, Akademiya Nauk USSR (Moscow)*, **130**, 104–13.

Shimek, S.J. and Monk, A. (1977) Daily activity of sea otters off the Monterey Peninsula, California. *Journal of Wildlife Management*, **41**, 277–83.

Shinitsky, M. (1984) *Physiology of Membrane fluidity*. Vol. 1. CRC Press Inc., Boca Raton, Forida.

Shinitsky, M. (1985) *Physiology of Membrane fluidity*. Vol. 2. CRC Press Inc., Boca Raton, Florida.

Singer, S.J. and Nicolson, G.L. (1972) The fluid mosiac model of the structure of cell membranes. *Science*, **175**, 720.

Smith, E.N. (1975) Thermoregulation of the American Alligator, *Alligator mississippiensis*. *Physiological Zoology*, **48**, 326–37.

Smith, E.N. (1979) Behavioural and physiological thermoregulation of crocodilians. *American Zoologist*, **19**, 239–47.

Smit-Vis, J.H. and Smit, G.J. (1963) Occurrence of hibernation in the golden hamster, *Mesocricetus auratus* Waterhouse. *Experientia*, **19**, 363–4.

Somme, L. (1964) Effects of glycerol on cold hardiness in insects. *Canadian Journal of Zoology*, **42**, 89–101.

Somme, L. (1967) The effect of temperature and anoxia on haemolymph composition and supercooling in three overwintering insects. *Journal of Insect Physiology*, **13**, 805–14.

Somme, L. (1982) Supercooling and winter survival in terrestrial arthropods. *Comparative Biochemistry and Physiology*, **73A**, 519–43.

Soper, T. (1982) *Birdwatch* Webb and Bower, Exeter.

Southwick, E.E. (1982) Metabolic energy of intact honey bee colonies. *Comparative Biochemistry and Physiology*, **71A**, 277–81.

Southwick, E.E. (1985) Allometric relations, metabolism and heat conductance in clusters of honey bees at cool temperatures. *Journal of Comparative Physiology B*, **156**, 143–9.

Spotila, J.R., Lommen, P.W., Bakken, G.S. and Gates, D.M. (1973) A mathematical model for body temperature of large reptiles: implications for dinosaur ecology. *American Naturalist*, **107**, 391–404.

Spradbery, J.P. (1973) *Wasps*. Sidgwick and Jackson, London.

Steen, J.B. and Gabrielsen, G.W. (1986) Thermogenesis in newly hatched eider (*Somateria mollissima*) and long-tailed (*Clangula hyemalis*) duckling and barnacle (*Branta leucopsis*) goslings. *Polar Research*, **4**, 81–6.

Steiner, J. and Grillmair, E. (1973) Possible galactic causes for periodic and episodic glaciations. *Bulletin of the Geological Society of America*, **84**, 1003–18.

Stern, C. (1926) An effect of temperature and age on crossing-over in the first chromosome of *Drosophila melanogaster*. *Proceedings of the National Academy of Sciences of the U.S.A.*, **12**, 530–2.

Stonehouse, B. (1953) The emperor penguin *Aptenodytes forsteri*. I. Breeding behaviour and development. *Scientific Reports of the Falkland Islands Dependency Survey*, **6**, 1–33.

Stonehouse, B. (1982) La zonation écologique sous les hautes latitudes australes. *Comité National Francais des Recherches Antarctiques*, **51**, 531–7.

Stonehouse, B. (1989) *Polar ecology*. Blackie, Glasgow and London.

Storey, K.B. (1985) Freeze tolerance in terrestrial frogs. *Cryo-Letters*, **6**, 115–34.

Storey, K.B. Baust, J.G. and Storey, J.M. (1981) Intermediary metabolism during low temperature acclimation in the overwintering gall fly larva *Eurosta solidaginis*. *Journal of Comparative Physiology*, **144**, 183–99.

Storey, K.B. and Storey, J.M. (1984) Biochemical adaption for freezing tolerance in the wood frog, *Rana sylvatica*. *Journal of Comparative Physiology B*, **155**, 29–36.

Storey, K.B. and Storey, J.M. (1986) Freeze tolerance and intolerance as strategies of winter survival in terrestrially-hibernating amphibians. *Comparative Biochemistry and Physiology*, **83A**, 613–7.

Storey, K.B. and Storey, J.M. (1988) Freeze tolerance: constraining forces, adaptive mechanisms. *Canadian Journal of Zoology*, **66**, 1122–7.

Sullivan, C.M. (1954) Temperature reception and responses in fish. *Journal of the Fisheries Research Board of Canada*, **11**, 153–70.

Sutcliffe, A.J. (1985) *On the track of ice age mammals*. British Museum (Natural History), London.

Swan, H. (1981) Neuroendocrine aspects of hibernation. In *Survival in the cold: hibernation and other adaptations*. (eds X.J. Musacchia and L. Jansky). Elsevier, New York.

Taylor, C.R. (1987) Structural and functional limits to oxidative metabolism: insights from scaling. *Annual Review of Physiology*, **49**, 135–46.

Templeman, W. (1948) The life history of the capelin (*Mallotus villosus*) (O.F. Müller) in Newfoundland waters. *Bulletin of the Newfoundland Government Laboratory*. No. 17, 151 pp.

Templeman, W. (1965) Mass mortalities of marine fishes in the Newfoundland area presumably due to low temperature. *Special Publications. International Commission for the Northwest Atlantic Fisheries*. No. 6. 137–47.

Theede, H. (1972) Vergleischende ökologisch-physiologische Untersuchungen zur zellären Kälterrestienz mariner evertebraten. *Marine Biology*, **15**, 160–91.

Theede, H., Schneppenheim, R. and Beress, L. (1976) Frostschutz-glykoproteine bei *Mytilus edulis*? *Marine Biology*, **36**, 183–9.

Thiriot-Quiévreux, C., Soyer, J., Bovee, F. de and Albert, P. (1988) Unusual complement in the brooding bivalve *Lasaea consanguinea*. *Genetica*, **76**, 143–51.

Thiriot-Quiévreux, C., Insua Pomba, A.M. and Albert, P. (1989) Polyploidie chez un bivalve incumbant, *Lasaea rubra* (Montagu). *Comptes Rendus. Academie des Sciences, Paris*, **308** III, 115–20.

Tøien, Ø, Aulie, A. and Steen, J.B. (1986) Thermoregulatory responses to egg cooling in incubating bantam hens. *Journal of Comparative Physiology B*, **156**, 303–7.

Trivedi, B. and Danforth, W.H. (1966) Effect of pH on the kinetics of frog muscle phosphofructokinase. *Journal of Biological Chemistry*, **241**, 4110–12.

Turnage, W.V. (1939) Desert subsoil temperatures. *Soil Science*, **47**, 195–9.

Turner, J.D., Schrag, J.D. and Devries, A.L. (1985) Ocular freezing avoidance in antarctic fish. *Journal of Experimental Biology*, **118**, 121–31.

Twente, J.W. and Twente, J.A. (1967) Seasonal variation in the hibernating behaviour of *Clitellus lateralis*. In *Mammalian hibernation*. Vol. 3. (eds K.C. Fisher, A.R. Dawe, C.P. Lyman, E. Schonbaum and F.E. Fouth). Oliver and Boyd, Edinburgh.

Tyler-Walters, H. and Crisp, D.J. (1989) The modes of reproduction in *Lasaea rubra* (Montagu) and *L. australis* (Lamarck): (Erycinidae; Bivalvia). In *Reproduction, genetics and distributions of marine organisms*. (eds J.S. Ryland and P.A. Tyler). Olsen and Olsen, Denmark.

Underwood, L.S. and Reynolds, P. (1980) Photoperiod and fur lengths in the Arctic fox (*Alopex lagopus* L.) *International Journal of Biometeorology*, **24**, 39–48.

Van't Hoff, T.H. (1884) *Etudes de dynamic chimique*. Muller, Amsterdam.

Van Voorhies, W.V., Raymond, J.A. and Devries A.L. (1978) Glycoproteins as biological antifreeze agents in the cod *Gadus ogac* (Richardson). *Physiological Zoology*, **51**, 347–53.

Vernberg, F.J. and Silverthorn, S.U. (1979) Temperature and osmoregulation in aquatic species. In *Mechanisms of Osmoregulation in Animals, Maintenance of Cell Volume*. (ed. R. Gilles) pp. 273–377. John Wiley and Sons, Chichester.

Viitanen, P. (1967) Hibernation and seasonal movements of the viper, *Vipera berus* (L.), in southern Finland. *Annales Zoologica Fennica*, **4**, 472–546.

Vogt, F.D. and Lynch, G.R. (1982) Influence of ambient temperature, nest availability, huddling and daily torpor on energy expenditure in the white-footed mouse *Peroymscus leucopus*. *Physiological Zoology*, **55**, 56–63.

Ward, M. (1975) *Mountain medicine: a clinical study of cold and high altitude*. St. Albans, New York.

White, M.J.D. (1973) *Animal cytology and evolution*. Cambridge University Press, Cambridge.

Williams, E.H. (1981) Thermal influences on oviposition in the montane butterfly *Euphrydates gilletti*. *Oecologia*, **50**, 342–6.

Williams, R.J. (1970) Freezing tolerance in *Mytilus edulis*. *Comparative Biochemistry and Physiology*, **35**, 145–61.

Wittingham, P. (1965) Problems of survival. In *Exploration medicine* (eds O.G. Edholm and A.L. Bacharach). John Wright & Sons, London.

Wohlschlag, D.E. (1964) Respiratory metabolism and ecological characteristics of some fishes in McMurdo Sound, Antarctica. *Antarctic Research Series* **1**, 33–62.

Wyatt, G.R. (1967) The biochemistry of sugars and polysaccharides in insects. *Journal of Insect Physiology*, **4**, 287–360.

Wyndham, C.H. and Loots, H. (1969) Responses to cold in Antarctica. *Journal of Applied Physiology*, **27**, 696.

Zachariassen, K.E. (1982) Nucleating agents in cold-hardy insects. *Comparative Biochemistry and Physiology*, **73A**, 557–62.

Zachariassen, K.E. (1985) Physiology of cold tolerance in insects. *Physiological Reviews*, **65**, 799–832.

Zachariassen, K.E. and Hammel, H.T. (1976) Nucleating agents in the haemolymph of insects tolerant to freezing. *Physiological Reviews*, **65**, 799–832.

Zeuthen, E. (1970) Rate of living as related to body size in organisms. *Polskie Archivum Hydrobiologii*, **17**, 21–30.

Zhegunov, G.F. (1987) Possible mechanism of reorganization of cardiomycete plasma membranes in hibernation-arousal-wake cycle of the ground squirrel *Citellus undulatus*. *Zhurnal Evoliutsionni Biokhimi i Fiziologii*, **23**, 647–51.

Zhegunov, G.F., Mikulinsky, Y.E. and Kudokotseva, E.V. (1988) Hyperactivation of protein synthesis in tissues of hibernating animals on arousal. *Cryo-letters*, **9**, 236–45.

Zhuchencho, A.A., Korol, A.B. and Kovyyukh, L.P. (1985) Change of the crossing over frequency in *Drosophila* during selection for resistance to temperature fluctuations. *Genetica*, **67**, 73–8.

Zwaan, A. De (1977) Anaerobic energy metabolism in bivalve molluscs. In *Annual Review of Oceangraphy and Marine Biology*, (ed. H. Barnes) **15**, 103–87.

Index